D1226403

SPACECRAFT

Name _____

Address _____

City _____ State ___ Zip ___

Put stamp here.
The Post Office
will not deliver
mail without
postage.

Krieger Publishing Company
P.O. Box 9542
Melbourne, FL 32902-9542

KRIEGER PUBLISHING COMPANY

MALABAR, FLORIDA
1991

Original Edition 1991

Printed and Published by
Krieger Publishing Company
Krieger Drive
Malabar, Florida 32950

Copyright © 1991 by Krieger Publishing Company

All rights reserved. No part of this book may be reproduced in any form
or by any means, electronic or mechanical, including storage and retrieval
systems without permission in writing from the publisher.

No liability is assumed with respect to the use of the information contained
herein.

Printed in the United States of America

Library of Congress Cataloging-in-Publication Data

Chobotov, Vladimir A.
    Spacecraft attitude dynamics and control/V. A. Chobotov. —
Original ed.
      p. 150 cm. — (Orbit, a foundation series)
    Includes bibliographical references and index.
    ISBN 0-89464-031-3 (acid-free paper)
    1. Astrodynamics. 2. Space vehicles—Attitude control systems.
I. Title
TL1050.C48   1991   90-28527
629.47′42—dc20   CIP

10 9 8 7 6 5 4 3 2

Series editor
Edwin F. Strother, Ph.D.

Contents

Introduction

This book is the outgrowth of the author's many years of professional activities in the field of spacecraft dynamics and orbital mechanics. It is also based on the courses taught at The Aerospace Corporation, Northrop University and the UCLA Extension/Department of Engineering. The book should therefore be a useful textbook for instruction at the university level or as a reference work for engineers engaged in research, design and development in this field. The book represents a comprehensive and self-contained treatment of the fundamentals of kinematics, rigid body dynamics, linear control theory, orbital environmental effects, and an introduction to the theory of the stability of motion. It can thus be regarded as a "road map" in the field of spacecraft dynamics and control.

Application of the theoretical developments is illustrated by numerous examples of actual spacecraft. Spin, dual-spin, three axis active, reaction wheel, control moment gyro, gravity gradient, and magnetic control systems are described. Topics such as active nutation damping (with gas jets and accelerometer sensing), separation dynamics of spinning bodies, and tethers in space (including deployment methods and elastic effects) are also considered. Environmental effects stemming from gravitation, solar radiation, aerodynamics, and geomagnetics are described in sufficient detail for understanding and practical application purposes. Thus, a good balance between theory and application has been achieved within a single volume treating many aspects of spacecraft dynamics and control.

Appendix A includes numerous exercises for student/teacher use with answers to problems provided where practical. The problems illustrate the theoretical concepts discussed in the book and are generally arranged in the order of difficulty with the confidence-building problems presented first.

Acknowledgements

I wish to acknowledge The Aerospace Corporation for the environment in which this book could be written. In particular, I am thankful to my colleagues Alan B. Jenkin and Dr. Daniel H. Platus for verifying derivations and making valuable suggestions in Chapter 2. Also, gratefully acknowledged are the reviews of Chapter 4 by Denny D. Pidhayny and of Chapter 7 by Dr. Thomas L. Alley.

The many valuable discussions and the assistance given by my son Dr. Michael V. Chobotov with computer graphics are greatly appreciated. Also, the patience, understanding and inspiration provided by my wife Lydia made the preparation of the manuscript a pleasant undertaking.

Special thanks go to Mary Roberts for expert copy editing, and to Professor Edwin F. Strother for a very thorough review and fine tuning of the manuscript at the Krieger Publishing Company in Melbourne, Florida.

Chapter 1
Kinematics and Dynamics of Angular Motion

1.1 Spacecraft Attitude Control Systems

One of the most important problems in spacecraft design is that of attitude stabilization and control. This is an extensive problem, as the missions of the space vehicles and their attitude requirements vary greatly. Also, the procedures used between the launch phase and on-orbit operations are somewhat complex. Consequently, the disciplines involved are numerous and include a combination of mathematics, dynamics, and control theory. In general, a spacecraft attitude control system consists of the following four major functional sections: sensing, logic, actuation, and vehicle dynamics. The sensing function determines satellite attitude. The logic programs the electronic signals in a correct sequence to the torque producing elements, which in turn rotate the spacecraft about its center of mass. The resulting motion (dynamics) is then monitored by the vehicle sensors which thus close the loop of the spacecraft attitude control system (see Figure 1.1).

The basic types of control systems are spin, three-axis active, and passive or gravity gradient control systems. A variation on spin control is the dual-spin system where a platform is "despun" from the rotating part of the spacecraft so that it can be pointed to any arbitrary direction in space. Three-axis active control may be achieved by mass expulsion or a combination of mass expulsion and a reaction wheel (RW) if the system contains a wheel or a set of wheels capable of spinning about an axis. The characteristics, advantages, and disadvantages of the various control systems are summarized in the following sections.

1.1.1 Spin Control

Spin stabilization provides simplicity, low cost, high reliability, and long system life. The entire spacecraft is spun, permitting fixed inertial orientation. Gyroscopic resistance provides stabilization about transverse axes. Thus, the spinning spacecraft possesses inherent resistance to external disturbance torques but is subject to nutation and precession dynamics. Sensor imaging is limited to line scanning ob-

tained by spin motion. Only one axis (spin) is oriented to a fixed reference which has poor maneuverability due to high angular momentum of the spacecraft. Ground assistance for momentum precession control is normally required. Power efficiency is inherently poor due to solar arrays being located on the spinning body since only half of the solar cells on a spinning satellite of the type shown in Figure 1.2 are illuminated by the sun continuously.

1.1.2 Dual-Spin Control

As in the case of a pure spinner, the major portion of the dual-spin controlled spacecraft is spun while only the payload section (platform) is despun. Autonomous platform pointing about the spin axis provides for continuous Earth viewing from the platform. Momentum stabilization about the transverse axes is similar to that of pure spin control systems and normally requires ground assistance for precession control. A primary advantage of this system is that fixed inertial orientation with scanning and pointing capability is possible. Also, spin about the longitudinal axis of the spacecraft is possible with energy dissipation on the platform. A disadvantage, however, is the somewhat limited growth potential due to the inefficient solar arrays on the spinner. Sensitivity to mass imbalances and nutational dynamics also adds to the cost and complexity of such systems (see Figure 1.3).

1.1.3 Three-Axis Active Control

In three-axis active control systems the major part of the spacecraft is despun. The payload is mounted on the main body which is controlled in attitude by either a mass expulsion subsystem (jets) or reaction wheels. Autonomous control about all axes can thus be provided for the configuration shown by Figure 1.4. A primary advantage of the system is that high pointing accuracy, which is limited only by that of the sensing subsystems, can be achieved. Three-axis control systems are adaptable to changing mission, maneuvering, and precision orientation requirements. The disadvantages are relatively costly hardware, greater weight (e.g., propellants), and power requirements. Also,

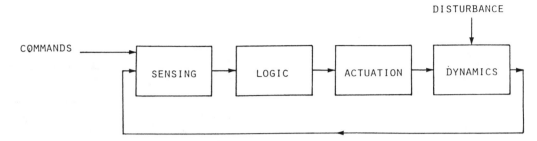

Figure 1.1   General satellite attitude control system.

Syncom satellite.

Figure 1.2   A typical spin stabilized spacecraft configuration.

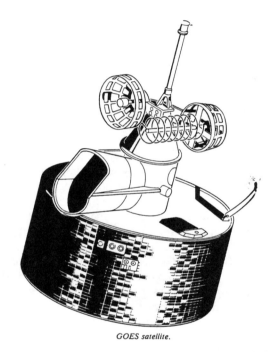

GOES satellite.

Figure 1.3   A typical spacecraft utilizing dual-spin stabilization.

momentum unloading (dumping) can occur when the RWs reach a limiting angular velocity. Complex thrust vector control and extensive fault detection schemes are often required for backup in three-axis control systems.

1.1.4   Momentum Bias Control

As in the three-axis control, the momentum bias system has the major portion of the spacecraft despun. The payload is mounted on the main body which controls a wheel spinning at a constant angular velocity normal to the orbital plane. Autonomous pitch (about the perpendicular to the orbital plane) and roll (about the velocity vector) pointing is provided. Yaw (about the local vertical) control is achieved with orbital coupling. Advantages of this system are the reduced weight and lower cost when compared to pure mass expulsion systems. This results from the momentum exchange capability which permits absorption of the cyclical disturbance torques by the RW. The orbit rate

coupling eliminates the need for a separate yaw sensor. A possible disadvantage of the system is a poor yaw accuracy capability. Also, the bias momentum wheel results in poor maneuverability of the system.

1.1.5   Passive (Gravity Gradient) Control

Long life and continuous Earth pointing are the primary advantages of a passive or gravity gradient control system. Simplicity of design, no mass or energy expenditure, and low cost are additional advantages of a gravity gradient control system. Two or three-axis control is possible depending on the inertia characteristics of the spacecraft. The main disadvantage is the poor pointing accuracy with respect to the orbiting reference frame. While Earth pointing accuracy to within 10 to 20 degrees is achievable with a purely passive system, pointing to within one degree is also possible with the aid of a small RW. The typically low restoring torques can be enhanced with the addition of

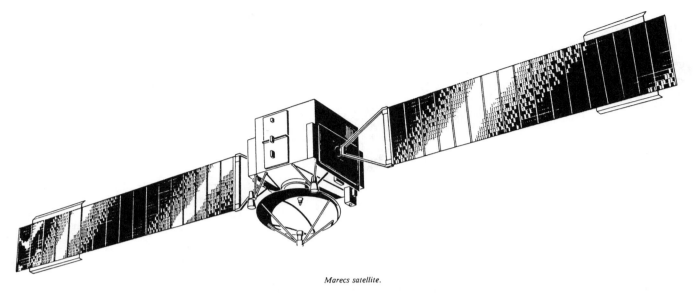

Marecs satellite.

Figure 1.4 An example of a spacecraft utilizing three-axis control.

extendable booms with tip masses. The masses increase the moment of inertia of the spacecraft and may contain the dampers necessary to limit the libration amplitudes of the satellite. A single boom gravity configuration is shown in Figure 1.5.

Eole satellite.

Figure 1.5 An example of a spacecraft utilizing gravity gradient stabilization.

1.1.6 Attitude Sensors

The most common sensors used in attitude control are the Earth horizon, sun, star, and inertial measuring units (IMUs). Magnetometers and radio interferometers are also used. Earth horizon sensors may be of the scanning or nonscanning type whose accuracy is in the tenth of a degree range. Sun and star sensors provide the highest accuracy in the arc second range. IMUs consist of accelerometers and gyros which provide high accuracy but require frequent updating due to instrument drift rates. The accuracy of the horizon sensors is limited by horizon definition and degrades at lower altitudes. Magnetometers can provide directional accuracies to within one degree of the magnetic field vector.

1.1.7 Control Actuators

The actuators or the torque-producing elements in the spacecraft attitude control system are generally a mass expulsion, momentum exchange, or environmental type. The mass expulsion actuators are the reaction control jets (using hot or cold gases). Reaction wheels (RWs) or control moment gyros (CMGs) are representative of the momentum exchange type. Magnetic, gravity gradient, and aerodynamic actuators are environmental types. Typical control torques are listed in Table 1.1. The results show that CMGs have the widest range of available torque authority. They can therefore be used to control the smallest and the largest satellites.

1.1.8 Control System Comparisons

The reference, accuracy, slew rate capability, payload efficiency, and design life are compared in Table 1.2 for the principal types of attitude control systems. The range of

Table 1.1. Control Actuator Torque Values

Torque Control System	Available Torque Range (newton · meters)
Reaction control (RCS)	10^{-2} to 10
Magnetic torquer	10^{-2} to 10^{-1}
Gravity gradient	10^{-6} to 10^{-3}
Aerodynamic	10^{-5} to 10^{-3}
Reaction wheel (RW)	10^{-1} to 1
Control moment gyro (CMG)	10^{-2} to 10^{3}

pointing accuracies is limited only by the sensor accuracies for all except the gravity gradient types. The capability to reorient rapidly (slew rate) is practically nonexistent for all but the three-axis control systems. The payload efficiency refers to the ability to point the antenna, a sensor, or an instrument in an arbitrary direction. Life expectancy is determined mostly by the amount of propellant required to perform the mission. Active three-axis systems are thus considered to be low life systems and passive, or gravity-gradient, spacecraft are long life (>10 yr) designs.

1.2 Basic Concepts of Kinematics

1.2.1 Velocity and Acceleration

The study of the kinematics and dynamics of motion requires a reference coordinate frame in which the position, velocity, and acceleration of a mass point can be specified. A nonrotating and nonaccelerating reference frame is known as an inertial frame in which the laws of mechanics are valid and can most conveniently be expressed. If an orthogonal inertial frame is used to denote the vector position $\vec{r}$ of a point, then its velocity and acceleration are $\vec{v} = d\vec{r}/dt$ and $\vec{a} = d\vec{v}/dt$, respectively. If the position $\vec{r}$ is measured in a frame which has an angular velocity $\vec{\omega}$ and the translational velocity $\vec{v}_o$, then the absolute or inertial velocity of the point is

$$\vec{v}_{\text{inertial}} = \dot{\vec{r}}' + \vec{\omega} \times \vec{r} + \vec{v}_o \qquad (1.1)$$

where $\dot{\vec{r}}'$ = translational velocity of the point measured in the rotating frame.

The acceleration of the point with respect to inertial space is

$$
\begin{aligned}
\vec{a} &= \frac{d\vec{v}}{dt} \\
&= \ddot{\vec{r}}' + \vec{\omega} \times \dot{\vec{r}}' + \dot{\vec{\omega}} \times \vec{r} + \vec{\omega} \\
&\quad \times (\dot{\vec{r}}' + \vec{\omega} \times \vec{r}) + \dot{\vec{v}}_o \\
&= \ddot{\vec{r}}' + 2\vec{\omega} \times \dot{\vec{r}}' + \dot{\vec{\omega}} \\
&\quad \times \vec{r} + \vec{\omega} \times (\vec{\omega} \times \vec{r}) + \vec{a}_o \qquad (1.2)
\end{aligned}
$$

where

$\ddot{\vec{r}}'$ = apparent (relative) acceleration of the point in the rotating frame

and

$\vec{a}_o$ = acceleration of the rotating frame.

The second and fourth terms in equation 1.2 are the Coriolis and centripetal accelerations, respectively. The $\dot{\vec{\omega}} \times \vec{r}$ term is the tangential acceleration. The Coriolis acceleration $2\vec{\omega} \times \dot{\vec{r}}'$ term exists only when there is a relative translational velocity $\dot{\vec{r}}'$ in a rotating frame. The Coriolis acceleration causes a particle moving along the Earth's surface (in a tangent plane) to drift to the right in the Northern Hemisphere and to the left in the Southern Hemisphere. This acceleration determines the sense of the vortex rotation of storms of the cyclone type. Examples of applications of the Coriolis acceleration can be found in references 1 and 2.

1.2.2 Euler Angles

The orientation of a rigid body with body fixed axes e_1, e_2, and e_3 associated with unit vectors $\hat{e}_1$, $\hat{e}_2$, and $\hat{e}_3$, respectively, can be specified in several ways relative to a reference frame E_1, E_2, and E_3 associated with unit vectors $\hat{E}_1$, $\hat{E}_2$, and $\hat{E}_3$. These include the classical Euler angles, sequential rotations about body axes, direction cosines, and the Eulerian parameters or "quaternions" as they are sometimes called. All these representations have certain advantages and limitations which depend on the type of motion to be described. The classical Euler angles, for example, are easier to visualize and are more convenient when working with spinning bodies. Sequential rotations are more advantageous when only small deviations from the reference

Table 1.2. A Comparison of Various Control Systems

Control System	Reference Orientation	Range of Orientation Accuracy (Deg)	Slew Rate Capability	Payload Efficiency	Life Expectancy (Years)
Spin	Sun/Earth Inertial	0.01 to 1.0	None	Low	7–10
Dual Spin	Sun/Earth Inertial	0.01 to 1.0	None	High	5–10
Three Axis	Sun/Earth Inertial	0.01 to 1.0	Arbitrary	High	3–7
Momentum Bias	Sun/Earth Inertial	0.01 to 1.0	None	High	5–15
Gravity Gradient	Earth Pointing	1 to 10	None	Low	>10

frame are involved. Such representations have the advantage that only three coordinates (angles) are required. The direction cosines, on the other hand, require certain constraint relations to reduce the number of independent parameters to three from the nine elements of the rotation matrix. Similarly, the quaternions have four parameters defining the orientation of the body relative to a reference frame. Their use, however, avoids the singularity of the Eulerian formulation resulting when one of the angles is zero, and facilitates numerical evaluation since only algebraic operations are involved.

1.2.3 Sequential Orthogonal Rotations

Consider the rotation of a unit orthogonal triad $\hat{e}_\alpha$ through an angle θ_1 relative to a reference unit triad $\hat{E}_\beta$ where the α and β subscripts take the values of 1, 2, 3 as shown in Figure 1.6.

The components of $\hat{E}_\beta$ along the $\hat{e}_\alpha$ directions are given by

$$\hat{e}_1 = \hat{E}_1$$
$$\hat{e}_2 = \hat{E}_2 \cos \theta_1 + \hat{E}_3 \sin \theta_1$$
$$\hat{e}_3 = -\hat{E}_2 \sin \theta_1 + \hat{E}_3 \cos \theta_1 \qquad (1.3)$$

or in matrix form as

$$\begin{pmatrix} \hat{e}_1 \\ \hat{e}_2 \\ \hat{e}_3 \end{pmatrix} = \begin{pmatrix} 1 & 0 & 0 \\ 0 & c\theta_1 & s\theta_1 \\ 0 & -s\theta_1 & c\theta_1 \end{pmatrix} \begin{pmatrix} \hat{E}_1 \\ \hat{E}_2 \\ \hat{E}_3 \end{pmatrix} = R(\theta_1) \begin{pmatrix} \hat{E}_1 \\ \hat{E}_2 \\ \hat{E}_3 \end{pmatrix} \qquad (1.4)$$

where $c\theta$ and $s\theta$ denote $\cos \theta$ and $\sin \theta$, respectively, and where the rotational matrix is

$$R(\theta_1) = \begin{pmatrix} 1 & 0 & 0 \\ 0 & c\theta_1 & s\theta_1 \\ 0 & -s\theta_1 & c\theta_1 \end{pmatrix}. \qquad (1.5)$$

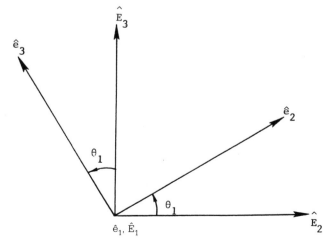

Figure 1.6 An example of an orthogonal rotation in two dimensions.

Equation 1.5 is an orthogonal rotation matrix which represents the rotation of the $\hat{e}_\alpha$ unit triad about the $\hat{E}_1$ unit vector. Similar rotation matrices may be used to represent rotations about other axes. For example, Figure 1.7 illustrates sequential θ_1, θ_2, θ_3 rotations of the $\hat{e}_\alpha$ unit triad, initially aligned with the $\hat{E}_\alpha$ unit triad, about the $\hat{E}_1$ (or $\hat{e}_1$) axis, the rotated $\hat{e}_2$ (or $\hat{e}_2'$) axis and, finally, the rotated $\hat{e}_3$ (or $\hat{e}_3''$) axis.

Each rotation may be expressed in terms of the orthogonal rotation matrices as follows:

$$\begin{pmatrix} \hat{e}_1' \\ \hat{e}_2' \\ \hat{e}_3' \end{pmatrix} = R(\theta_1) \begin{pmatrix} \hat{E}_1 \\ \hat{E}_2 \\ \hat{E}_3 \end{pmatrix} \qquad (1.6)$$

$$\begin{pmatrix} \hat{e}_1'' \\ \hat{e}_2'' \\ \hat{e}_3'' \end{pmatrix} = R(\theta_2) \begin{pmatrix} \hat{e}_1' \\ \hat{e}_2' \\ \hat{e}_3' \end{pmatrix} \qquad (1.7)$$

$$\begin{pmatrix} \hat{e}_1 \\ \hat{e}_2 \\ \hat{e}_3 \end{pmatrix} = R(\theta_3) \begin{pmatrix} \hat{e}_1'' \\ \hat{e}_2'' \\ \hat{e}_3'' \end{pmatrix} \qquad (1.8)$$

$$\begin{pmatrix} \hat{e}_1 \\ \hat{e}_2 \\ \hat{e}_3 \end{pmatrix} = R(\theta_3) R(\theta_2) R(\theta_1) \begin{pmatrix} \hat{E}_1 \\ \hat{E}_2 \\ \hat{E}_3 \end{pmatrix} \qquad (1.9)$$

where

$$R(\theta_1) = \begin{pmatrix} 1 & 0 & 0 \\ 0 & c\theta_1 & s\theta_1 \\ 0 & -s\theta_1 & c\theta_1 \end{pmatrix}, \quad R(\theta_2) = \begin{pmatrix} c\theta_2 & 0 & -s\theta_2 \\ 0 & 1 & 0 \\ s\theta_2 & 0 & c\theta_2 \end{pmatrix},$$

$$R(\theta_3) = \begin{pmatrix} c\theta_3 & s\theta_3 & 0 \\ -s\theta_3 & c\theta_3 & 0 \\ 0 & 0 & 1 \end{pmatrix}.$$

The combined matrix for the 1, 2, 3 sequence of rotation is the product of the orthogonal rotation matrices. It is of the form

$$R_{123} = R(\theta_3) R(\theta_2) R(\theta_1)$$

$$= \qquad (1.10)$$

$$\begin{pmatrix} c\theta_2 c\theta_3 & c\theta_3 s\theta_1 s\theta_2 + c\theta_1 s\theta_3 & -c\theta_1 c\theta_3 s\theta_2 + s\theta_1 s\theta_3 \\ -c\theta_2 s\theta_3 & -s\theta_1 s\theta_2 s\theta_3 + c\theta_1 c\theta_3 & c\theta_1 s\theta_2 s\theta_3 + s\theta_1 c\theta_3 \\ s\theta_2 & -c\theta_2 s\theta_1 & c\theta_1 c\theta_2 \end{pmatrix}$$

The inverse relationship in equation 1.9 is of the form

$$\begin{pmatrix} \hat{E}_1 \\ \hat{E}_2 \\ \hat{E}_3 \end{pmatrix} = R_{123}^{-1} \begin{pmatrix} \hat{e}_1 \\ \hat{e}_2 \\ \hat{e}_3 \end{pmatrix} \qquad (1.11)$$

where, because of orthogonality, the inverse matrix is equivalent to its transpose; that is,

$$R_{123}^{-1} = R_{123}'. \qquad (1.12)$$

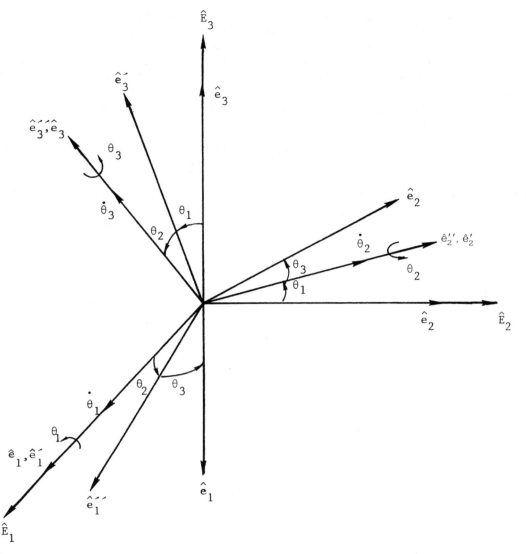

Figure 1.7 Sequential orthogonal rotations of the e_α reference frame about the E_α reference frame.

The inverse of the rotation matrix in equation 1.12 is obtained simply by interchanging the rows and columns of the matrix.

The orthogonal rotation matrices represent respective reference frames in terms of the rotation angles θ_i. Thus, for example, the components a_1, a_2, a_3 of a vector specified in the E_α frame are obtained in the e_α frame by evaluating the rotation matrix in equation 1.10 at the rotation angles θ_1, θ_2, and θ_3. Conversely, the components of a vector specified in the e_α frame are obtained in the E_α frame by the inverse relation found in equation 1.11.

1.2.3.1 Classical Euler Sequence of Rotations

The classical Euler sequence of rotations first involves a rotation about the $\hat{e}_3$ axis, then one about the rotated $\hat{e}_1$ axis, and finally a rotation about the $\hat{e}_3$ axis. This 3, 1, 3 sequence of rotations is illustrated in Figure 1.8 in terms of the Euler

angles ψ, θ, and φ. The associated Euler angle rates are $\dot{\psi}$, $\dot{\theta}$, and $\dot{\varphi}$. The sequence is particularly convenient for representing the orientation of spinning bodies, such as spacecraft, as will be shown later.

The rotation matrix for this case is of the form

$$\begin{pmatrix} \hat{e}_1 \\ \hat{e}_2 \\ \hat{e}_3 \end{pmatrix} = \begin{pmatrix} c\varphi & s\varphi & 0 \\ -s\varphi & c\varphi & 0 \\ 0 & 0 & 1 \end{pmatrix} \begin{pmatrix} 1 & 0 & 0 \\ 0 & c\theta & s\theta \\ 0 & -s\theta & c\theta \end{pmatrix}$$

$$\begin{pmatrix} c\psi & s\psi & 0 \\ -s\psi & c\psi & 0 \\ 0 & 0 & 1 \end{pmatrix} \begin{pmatrix} \hat{E}_1 \\ \hat{E}_2 \\ \hat{E}_3 \end{pmatrix}$$

$$= R_{313} \begin{pmatrix} \hat{E}_1 \\ \hat{E}_2 \\ \hat{E}_3 \end{pmatrix} \qquad (1.13)$$

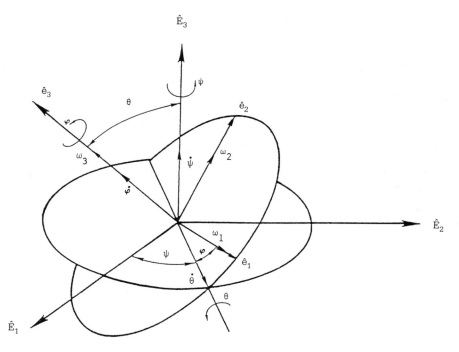

Figure 1.8 Classical Euler rotations of a rigid body.

where

$$R_{313} = \qquad (1.14)$$

$$\begin{pmatrix} c\varphi c\psi - s\varphi c\theta s\psi & c\varphi s\psi + s\varphi c\theta c\psi & s\varphi s\theta \\ -s\varphi c\psi - c\varphi c\theta s\psi & -s\varphi s\psi + c\varphi c\theta c\psi & c\varphi s\theta \\ s\theta s\psi & -s\theta c\psi & c\theta \end{pmatrix}$$

1.2.3.2 Euler Rates

The components of the Euler rates $\dot{\psi}$, $\dot{\theta}$, and $\dot{\varphi}$ along the e_α body axes can be obtained from Figure 1.8 in the following form [1]:

$$\omega_1 = \dot{\theta} \cos \varphi + \dot{\psi} \sin \theta \sin \varphi$$
$$\omega_2 = \dot{\psi} \sin \theta \cos \varphi - \dot{\theta} \sin \varphi$$
$$\omega_3 = \dot{\varphi} + \dot{\psi} \cos \theta \qquad (1.15)$$

These equations can be solved for $\dot{\psi}$, $\dot{\varphi}$, $\dot{\theta}$ as

$$\begin{pmatrix} \dot{\psi} \\ \dot{\varphi} \\ \dot{\theta} \end{pmatrix} = \frac{1}{\sin \theta} \times$$

$$\begin{pmatrix} \sin \varphi & \cos \varphi & 0 \\ -\sin \varphi \cos \theta & -\cos \varphi \cos \theta & \sin \theta \\ \cos \varphi \sin \theta & -\sin \varphi \sin \theta & 0 \end{pmatrix} \begin{pmatrix} \omega_1 \\ \omega_2 \\ \omega_3 \end{pmatrix}. \qquad (1.16)$$

It should be noted that equation 1.16 is a nonorthogonal matrix which has a singularity at $\theta = 0$.

1.2.4 Direction Cosines

The $\hat{e}_\alpha$ and $\hat{E}_\beta$ unit vector triads (where α, β subscripts range from 1 to 3) are related as follows:

$$\begin{pmatrix} \hat{e}_1 \\ \hat{e}_2 \\ \hat{e}_3 \end{pmatrix} = R \begin{pmatrix} \hat{E}_1 \\ \hat{E}_2 \\ \hat{E}_3 \end{pmatrix}, \qquad (1.17)$$

where R is a 3×3 rotation matrix of the form

$$R = \begin{pmatrix} a_{11} & a_{12} & a_{13} \\ a_{21} & a_{22} & a_{23} \\ a_{31} & a_{32} & a_{33} \end{pmatrix}. \qquad (1.18)$$

The elements $a_{\alpha\beta}$ of the R matrix are the direction cosines of the respective unit vectors between the $\hat{e}_\alpha$ and $\hat{E}_\beta$ triads. It is therefore necessary only to determine these elements to be able to define the orientation of a body reference frame relative to another reference frame. Using the direction cosines is advantageous when there are forces or torques which can be most easily expressed in a body fixed reference frame. Visualization of the body's orientation is difficult, however, because now there are nine elements to consider instead of three angles.

Unit vector triads $\hat{e}_\alpha$ and $\hat{E}_\beta$ (α, $\beta = 1 - 3$) can be related as

$$\hat{e}_\alpha = a_{\alpha\beta} \hat{E}_\beta \qquad (1.19)$$

where $a_{\alpha\beta}$ are the elements of R and the usual tensor summation convention for repeated indices is used. For example,

$$\hat{e}_1 = a_{11}\hat{E}_1 + a_{12}\hat{E}_2 + a_{13}\hat{E}_3 .$$

Because of the orthogonality of the rotation matrix

$$\hat{E}_\beta = a_{\alpha\beta}\hat{e}_a \qquad (1.20)$$

or, for example, summing over the α index for $\beta = 1$ yields

$$\hat{E}_1 = a_{11}\hat{e}_1 + a_{21}\hat{e}_2 + a_{31}\hat{e}_3 .$$

Also, it follows that

$$a_{\alpha\beta} = \hat{e}_\alpha \cdot \hat{E}_\beta . \qquad (1.21)$$

Some properties of the direction cosines $a_{\alpha\beta}$, which will later prove to be useful, are now presented.

1. The sum of the squares of the elements of a row (column) is equal to unity; for example,

$$a_{11}^2 + a_{21}^2 + a_{31}^2 = 1 \qquad (1.22)$$

2. The sum of the multiples of paired elements of two rows (columns) is equal to zero; for example,

$$a_{11}a_{21} + a_{12}a_{22} + a_{13}a_{23} = 0 \qquad (1.23)$$

3. Each element of the matrix is equal to its minor; for example,

$$a_{12} = -(a_{21}a_{33} - a_{23}a_{31}) \qquad (1.24)$$

1.2.4.1 Kinematical Equations of Poisson

If a body reference frame is undergoing a rotation, then the elements of the rotation matrix will be functions of time. The functional relationships between the direction cosines and their rates are known as the kinematical equations of Poisson [3]. They are of the form

$$\dot{a}_{11} = a_{12}\omega_3 - a_{13}\omega_2$$
$$\dot{a}_{12} = a_{13}\omega_1 - a_{11}\omega_3$$
$$\dot{a}_{13} = a_{11}\omega_2 - a_{12}\omega_1$$
$$\dot{a}_{21} = a_{22}\omega_3 - a_{23}\omega_2$$
$$\dot{a}_{22} = a_{23}\omega_1 - a_{21}\omega_3$$
$$\dot{a}_{23} = a_{21}\omega_2 - a_{22}\omega_1$$
$$\dot{a}_{31} = a_{32}\omega_3 - a_{33}\omega_2$$
$$\dot{a}_{32} = a_{33}\omega_1 - a_{31}\omega_3$$
$$\dot{a}_{33} = a_{31}\omega_2 - a_{32}\omega_1 \qquad (1.25)$$

Integration of the previous equations can be performed when the components of the body angular velocity $\vec{\omega}$ are given or can be obtained by solving Euler's dynamical equations. The resultant direction cosines then define the orientation of the body reference frame in terms of the rotation matrix as in equation 1.13, for example.

1.2.5 Quaternions

The attitude determination of rigid bodies by the use of the quaternion parameters has several advantages over the use of Euler angles or direction cosines. Quaternions involve the use of algebraic relations to determine the elements of the rotation matrix instead of trigonometric functions. The computations are faster and there are no singularities as

may occur in the Euler angle formulation. Moreover, fewer multiplications are required for propagating successive incremental rotations. A disadvantage is that one of the four components is redundant, and that there is, in general, no obvious physical interpretation of the rotation geometry [4, 5, 7, 9, 10].

A quaternion is a scalar plus a vector, totaling four elements. Three of the elements describe a vector which defines an axis of rotation. The fourth element, a scalar, defines the magnitude of a rotation angle about the axis of rotation. The formulation is based on Euler's theorem which states that any rotation of a body (or coordinate system) with respect to another may be described by a single rotation through some angle about a single fixed axis.

Let a quaternion $\vec{Q}$ be expressed as

$$\vec{Q} = \vec{q} + q_4$$
$$= iq_1 + jq_2 + kq_3 + q_4 \qquad (1.26)$$

where i, j, k are hyperimaginary numbers defined by Hamilton [9] and q_4 is a scalar. The $q_i(i = 1 - 4)$ are the Euler symmetric parameters or quaternions in short. They can be defined as

$$q_1 = m_1 \sin \mu/2$$
$$q_2 = m_2 \sin \mu/2$$
$$q_3 = m_3 \sin \mu/2$$
$$q_4 = \cos \mu/2 \qquad (1.27)$$

where m_1, m_2, m_3 are direction cosines of the Euler vector $\vec{q}$, and μ is the rotation angle about $\vec{q}$ as is illustrated in Figure 1.9.

The Euler vector $\vec{q}$ is defined as

$$\vec{q} = \hat{\lambda} \sin \mu/2 \qquad (1.28)$$

where the unit vector $\hat{\lambda}$ is given as

$$\hat{\lambda} = \frac{q_1\hat{E}_1 + q_2\hat{E}_2 + q_3\hat{E}_3}{(q_1^2 + q_2^2 + q_3^2)^{1/2}} . \qquad (1.29)$$

Consider the relationship between a body fixed unit vector triad $\hat{e}_\alpha(\alpha = 1, 2, 3)$ and a reference unit vector triad $\hat{E}_\beta$ given in cartesian tensor notation as [3, 11]

$$\hat{e}_\alpha = a_{\alpha\beta}\hat{E}_\beta \qquad (1.30)$$

The elements $a_{\alpha\beta}$ are then given by the relationship

$$a_{\alpha\beta} = 2q_\alpha q_\beta + \delta_{\alpha\beta}(1 - q_\gamma q_\gamma) + 2\varepsilon_{\alpha\beta\gamma}q_\gamma q_4 \qquad (1.31)$$

where, $\delta_{\alpha\beta}$ is the Kronecker delta which is equal to unity when $\alpha = \beta$ and is zero otherwise.

Also, $\varepsilon_{\alpha\beta\gamma}$ is the three dimensional epsilon permutation symbol defined by

$$\varepsilon_{\alpha\beta\gamma} = \begin{cases} 1 \text{ for } \alpha, \beta, \gamma \text{ an even permutation of } 1, 2, 3 \\ -1 \text{ for } \alpha, \beta \ \gamma \text{ an odd permutation of } 1, 2, 3 \\ 0 \text{ otherwise (i.e., if any repetitions occur)} \end{cases}$$

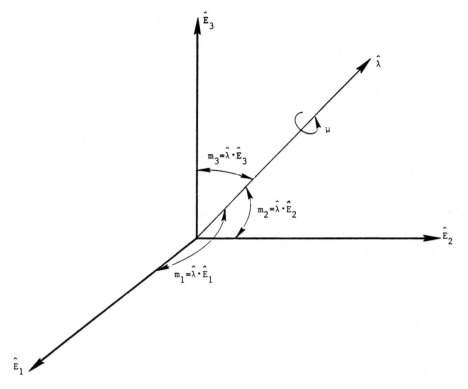

Figure 1.9 Euler vector orientation.

$$= (\hat{e}_\alpha \times \hat{e}_\beta) \cdot \hat{e}_\gamma$$

$$= \frac{1}{2}(\alpha - \beta)(\beta - \gamma)(\gamma - \alpha) . \qquad (1.32)$$

The α, β, γ indices range from 1 to 3.

In expanded form, equation (1.31) becomes

$$a_{11} = q_1^2 - q_2^2 - q_3^2 + q_4^2$$
$$a_{12} = 2(q_1 q_2 + q_3 q_4)$$
$$a_{13} = 2(q_1 q_3 - q_2 q_4)$$
$$a_{21} = 2(q_1 q_2 - q_3 q_4)$$
$$a_{22} = -q_1^2 + q_2^2 - q_3^2 + q_4^2$$
$$a_{23} = 2(q_1 q_4 + q_2 q_3)$$
$$a_{31} = 2(q_1 q_3 + q_2 q_4)$$
$$a_{32} = 2(-q_1 q_4 + q_2 q_3)$$
$$a_{33} = -q_1^2 - q_2^2 + q_3^2 + q_4^2 . \qquad (1.33)$$

The condition

$$q_i q_i = 1 \qquad (i = 1 - 4) \quad (1.34)$$

or

$$q_1^2 + q_2^2 + q_3^2 + q_4^2 = 1 \qquad (1.35)$$

is automatically satisfied and can be used for numerical control of machine computations.

In terms of the components of the rotation matrix $a_{\alpha\beta}$, the quaternions can be expressed as

$$q_1 = \frac{1}{4q_4}(a_{23} - a_{32})$$

$$q_2 = \frac{1}{4q_4}(a_{31} - a_{13}) \cdot$$

$$q_3 = \frac{1}{4q_4}(a_{12} - a_{21})$$

$$q_4 = \pm\frac{1}{2}(1 + a_{11} + a_{22} + a_{33})^{1/2} . \qquad (1.36)$$

The relationship between the quaternions q_i and the body angular velocities $\omega_\alpha(\alpha = 1, 2, 3)$ relative to the E_β frame is given by

$$\omega_\alpha = 2(\varepsilon_{\alpha\beta\gamma}\dot{q}_\beta q_\gamma + \dot{q}_\alpha q_4 - q_\alpha \dot{q}_4) \qquad (1.37)$$

or

$$\omega_i = 2q_{ij}\dot{q}_j \qquad (1.38)$$

where q_{ij} are the elements of the following four dimensional orthogonal rotation matrix:

$$q_{ij} = \begin{pmatrix} q_4 & q_3 & -q_2 & -q_1 \\ -q_3 & q_4 & q_1 & -q_2 \\ q_2 & -q_1 & q_4 & -q_3 \\ q_1 & q_2 & q_3 & q_4 \end{pmatrix} . \qquad (1.39)$$

Here $\dot{q}_i$ denotes the time derivative of the quaternion; namely,

$$\dot{q}_i = \frac{d}{dt}(q_i) . \qquad (1.40)$$

As before, the α, β, γ indices range from 1 to 3 while the values of i, j go from 1 to 4 inclusive. The ω_4 term in equation 1.38 is zero because of the identity equation 1.35. That is

$$\omega_4 = 2q_{4j}\dot{q}_j$$
$$= \frac{d}{dt}(q_i q_i) \qquad (1.41)$$
$$\overset{\Delta}{\equiv} 0 .$$

Expanded, equation 1.38 becomes

$$\omega_1 = 2(q_4\dot{q}_1 + q_3\dot{q}_2 - q_2\dot{q}_3 - q_1\dot{q}_4)$$
$$\omega_2 = 2(-q_3\dot{q}_1 + q_4\dot{q}_2 + q_1\dot{q}_3 - q_2\dot{q}_4)$$
$$\omega_3 = 2(q_2\dot{q}_1 - q_1\dot{q}_2 + q_4\dot{q}_3 - q_3\dot{q}_4) . \qquad (1.42)$$

The matrix q_{ij} in equation 1.39 is an orthogonal matrix in view of the identity (equation 1.34). It satisfies the orthogonality condition

$$q_{ij}q_{ik} = \delta_{jk} \qquad (1.43)$$

which can be verified by inspection. Equation 1.38 can be solved for $\dot{q}_j$ as follows [6]:

Multiplying equation 1.38 by q_{ik} yields

$$q_{ik}\omega_i = 2q_{ik}q_{ij}\dot{q}_j$$
$$= 2\delta_{kj}\dot{q}_j$$

or

$$\dot{q}_j = \frac{1}{2}q_{ij}\omega_i . \qquad (1.44)$$

Expanded, equation 1.44 becomes

$$\begin{pmatrix} \dot{q}_1 \\ \dot{q}_2 \\ \dot{q}_3 \\ \dot{q}_4 \end{pmatrix} = 1/2 \begin{pmatrix} q_4 & -q_3 & q_2 & q_2 \\ q_3 & q_4 & -q_1 & q_2 \\ -q_2 & q_1 & +q_4 & q_3 \\ -q_1 & -q_2 & -q_3 & q_4 \end{pmatrix} \begin{pmatrix} \omega_1 \\ \omega_2 \\ \omega_3 \\ 0 \end{pmatrix} \qquad (1.45)$$

and since the terms are linear in ω_1, ω_2, ω_3 and are also linear in q_j, this equation can also be written as

$$\begin{pmatrix} \dot{q}_1 \\ \dot{q}_2 \\ \dot{q}_3 \\ \dot{q}_4 \end{pmatrix} = 1/2 \begin{pmatrix} 0 & \omega_3 & -\omega_2 & \omega_1 \\ -\omega_3 & 0 & \omega_1 & \omega_2 \\ \omega_2 & -\omega_1 & 0 & \omega_3 \\ -\omega_1 & -\omega_2 & -\omega_3 & 0 \end{pmatrix} \begin{pmatrix} q_1 \\ q_2 \\ q_3 \\ q_4 \end{pmatrix} . \qquad (1.46)$$

It is worth noting that for small angular rotations θ_i ($i = 1 - 3$) about the respective unit vector directions $\hat{e}_i$, the quaternions become

$$q_i \approx \theta_i/2 \quad \text{and} \quad q_4 \approx 1 . \qquad (1.47)$$

The use of the quaternions is especially attractive in computer simulations where only algebraic operations involving quaternions and the direction cosines $a_{\alpha\beta}$ are involved. There is also an advantage in that the quaternions define the rotation unambiguously up to two revolutions of the body. The direction cosines, on the other hand, cannot determine whether the body has actually performed a complete revolution or is still in its original position.

1.2.6 Combined Rotations with Quaternions

Given a rotation matrix R (R may be a 3×3 matrix of Euler angles or direction cosines), one coordinate frame (e.g., body frame e_α) may be related to another frame (e.g., inertial reference E_β) as

$$\hat{e}_\alpha = R\hat{E}_\beta . \qquad (1.48)$$

For a subsequent (another) rotation of $\hat{e}_\alpha$ through a rotation matrix P, the original and final orientations of $\hat{e}_\alpha$ can be expressed as $\hat{e}'_\alpha$, where

$$\hat{e}'_\alpha = PR\hat{E}_\beta . \qquad (1.49)$$

Note that $\hat{e}'_\alpha = f[(\theta, \psi, \phi, \theta', \psi', \phi')]$ is a function of six rotation angles requiring 27 multiplications.

Quaternions can reduce the number of multiplications and hence the time required. Let the quaternion that describes the rotation of system A (reference axes) to system B be expressed as

$$Q_A^B = q_1, q_2, q_3, q_4 \qquad (1.50)$$

and another quaternion that describes the rotation of system B to system C as

$$Q_B^C = q'_1, q'_2, q'_3, q'_4 . \qquad (1.51)$$

Then

$$Q_A^C = q''_1, q''_2, q''_3, q''_4 \qquad (1.52)$$

where

$$Q_A^C = \begin{pmatrix} q'_4 & q'_3 & -q'_2 & q'_1 \\ -q'_3 & q'_4 & q'_1 & q'_2 \\ q'_2 & -q'_1 & q'_4 & q'_3 \\ -q'_1 & -q'_2 & -q'_3 & q'_4 \end{pmatrix} \begin{pmatrix} q_1 \\ q_2 \\ q_3 \\ q_4 \end{pmatrix} . \qquad (1.53)$$

Thus, only 16 multiplications, instead of the previous 27, are required for two sequential rotations.

1.3 Fundamentals of Dynamics

1.3.1 Angular Momentum and Kinetic Energy of a Body

The moment of momentum or the angular momentum of a rigid body is an important concept which can be developed as follows from basic principles. Consider a rigid body with body fixed axes $e_\alpha (\alpha = 1 - 3)$ rotating with an angular velocity $\vec{\omega}$ as illustrated in Figure 1.10. The angular momentum of an element of mass dm with velocity $\vec{v}$ is

$$d\vec{h} = \vec{r} \times \vec{v}\, dm \qquad (1.54)$$

where $\vec{r}$ is the position vector to dm from the origin of the coordinate frame e_α, which is coincident with the center of mass of the rigid body, and E_α, is an inertial reference frame.

Integrating equation 1.54 over the entire body, denoted here as B, the angular momentum is of the form

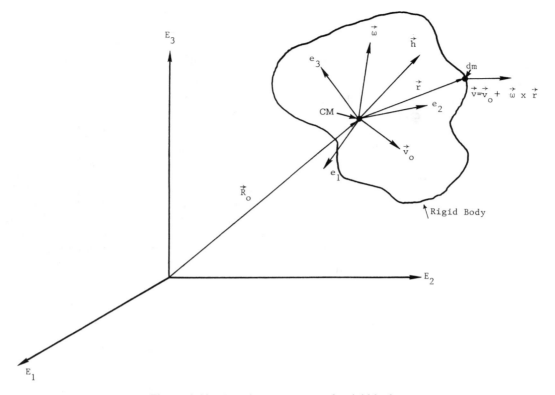

Figure 1.10   Angular momentum of a rigid body.

$$\vec{h} = \int_B \vec{r} \times \vec{v}\ dm$$

$$= \int_B \vec{r} \times (\vec{v}_o + \dot{\vec{r}}\,' + \vec{\omega} \times \vec{r})\ dm$$

$$= \int_B \vec{r} \times (\vec{\omega} \times \vec{r})\ dm \qquad (1.55)$$

since the integral of $\vec{r} \times \vec{v}_o$ over the body vanishes and $\dot{\vec{r}}\,'$ is zero for a rigid body.

Using the triple vector product identity

$$\vec{A} \times (\vec{B} \times \vec{C}) = (\vec{A} \cdot \vec{C})\vec{B} - (\vec{A} \cdot \vec{B})\vec{C}$$

the integral in equation 1.55 can be expressed in the form

$$\vec{h} = \int_B [(\vec{r} \cdot \vec{r})\vec{\omega} - (\vec{r} \cdot \vec{\omega})\vec{r}]\ dm$$

$$= \int_B (r^2\vec{\omega} - \vec{r}\ \vec{r} \cdot \vec{\omega})\ dm$$

$$= \int_B (r^2\bar{\bar{E}} - \vec{r}\ \vec{r})\ dm \cdot \vec{\omega}$$

$$= \bar{\bar{I}} \cdot \vec{\omega} \qquad (1.56)$$

where the inertia dyadic

$$\bar{\bar{I}} = \int_B (r^2\bar{\bar{E}} - \vec{r}\ \vec{r})\ dm \qquad (1.57)$$

and

$$\bar{\bar{E}} = \hat{e}_1\hat{e}_1 + \hat{e}_2\hat{e}_2 + \hat{e}_3\hat{e}_3 \qquad (1.58)$$

is a unit dyadic.

Letting $\vec{r} = r_\alpha\hat{e}_\alpha$, where r_α are the components of the position vector $\vec{r}$ along the e_α body axes (with unit vectors $\hat{e}_\alpha$) and, similarly, $\vec{\omega} = \omega_\alpha\hat{e}_\alpha$, the inertia dyadic $\bar{\bar{I}}$ can be expressed in the form $I = \bar{\bar{I}}_{\alpha\beta}\hat{e}_\alpha\hat{e}_\beta$ or expanded as

$$\bar{\bar{I}} = \begin{pmatrix} I_{11}\hat{e}_1\hat{e}_1 & I_{12}\hat{e}_1\hat{e}_2 & I_{13}\hat{e}_1\hat{e}_3 \\ I_{21}\hat{e}_2\hat{e}_1 & I_{22}\hat{e}_2\hat{e}_2 & I_{23}\hat{e}_2\hat{e}_3 \\ I_{31}\hat{e}_3\hat{e}_1 & I_{32}\hat{e}_3\hat{e}_2 & I_{33}\hat{e}_3\hat{e}_3 \end{pmatrix}. \qquad (1.59)$$

The body moments and products of inertia are calculated as follows [1]:

$$I_{11} = \int_B (r_2^2 + r_3^2)\ dm$$

$$I_{22} = \int_B (r_1^2 + r_3^2)\ dm$$

$$I_{33} = \int_B (r_1^2 + r_2^2)\ dm \qquad (1.60)$$

$$I_{12} = -\int_B r_1 r_2\ dm = I_{21}$$

$$I_{13} = -\int_B r_1 r_3\ dm = I_{31}$$

$$I_{23} = -\int_B r_2 r_3\ dm = I_{32}$$

The inertia matrix of the body is thus defined by the scalar terms in the inertia dyadic expression in equation 1.59.

1.3.2 Principal Moments of Inertia and Axes

The moments and products of inertia of a body evaluated from the integrals in equation 1.60 constitute the general inertia matrix for the body. If the body fixed reference frame e_α is chosen such that the off-diagonal terms in the inertia matrix are all zero [i.e., $I_{\alpha\beta} = 0 (\alpha \neq \beta)$], then the reference axes are the principal axes of the body, and the corresponding moments of inertia are the principal moments of inertia. The inertia dyadic can then be expressed as

$$\bar{I} = I_\alpha \hat{e}_\alpha \hat{e}_\alpha \qquad (\alpha = 1 - 3)$$
$$= I_1 \hat{e}_1 \hat{e}_1 + I_2 \hat{e}_2 \hat{e}_2 + I_3 \hat{e}_3 \hat{e}_3 \qquad (1.61)$$

where I_α terms are the principal moments of inertia of the body.

Equation 1.56 shows that the angular momentum vector $\vec{h}$ and the angular velocity vector $\vec{\omega}$ are, in general, not colinear. They are colinear only when the rotation of the body is about a principal axis. Then the angular momentum reduces to the particularly simple form

$$\vec{h} = I\vec{\omega} \qquad (1.62)$$

where I is the constant of proportionality. Thus, for example, if the rotation is about the principal axis e_1 (with a unit vector $\hat{e}_1$), then

$$\vec{\omega} = \omega_1 \hat{e}_1$$

and

$$\vec{h} = I\omega_1 \hat{e}_1$$
$$= h_1 \hat{e}_1 . \qquad (1.63)$$

For the general case of rotation about nonprincipal axes

$$\vec{h} = \bar{I} \cdot \vec{\omega}$$
$$= I_{\alpha\beta} \hat{e}_\alpha \hat{e}_\beta \cdot \omega_\beta \hat{e}_\beta$$
$$= I_{\alpha\beta} \omega_\beta \hat{e}_\alpha$$
$$= h_\alpha \hat{e}_\alpha . \qquad (1.64)$$

Here the e_1 axis component is found by letting $\alpha = 1$ and summing over β from 1 to 3 which yields

$$h_1 \hat{e}_1 = (I_{11}\omega_1 + I_{12}\omega_2 + I_{13}\omega_3)\hat{e}_1 . \qquad (1.65)$$

Equating the h_1 components in equations 1.64 and 1.65, and similarly for the h_2 and h_3 components yields

$$h_1 = I\omega_1 = I_{11}\omega_1 + I_{12}\omega_2 + I_{13}\omega_3$$
$$h_2 = I\omega_2 = I_{21}\omega_1 + I_{22}\omega_2 + I_{23}\omega_3 \qquad (1.66)$$
$$h_3 = I\omega_3 = I_{31}\omega_1 + I_{32}\omega_2 + I_{33}\omega_3$$

which can, in turn, be written in matrix form as

$$\begin{pmatrix} (I_{11} - I) & I_{12} & I_{13} \\ I_{21} & (I_{22} - I) & I_{23} \\ I_{31} & I_{32} & (I_{33} - I) \end{pmatrix} \begin{pmatrix} \omega_1 \\ \omega_2 \\ \omega_3 \end{pmatrix} = 0 \qquad (1.67)$$

Since ω_1, ω_2, ω_3 are not all zero, the inertia matrix in equation 1.67 represents the characteristic equation (a cubic in I) which can be solved for three real positive roots or eigenvalues (the principal moments of inertia I_1, I_2, I_3).

Principal axes are found from the corresponding ω_1/ω_2, ω_1/ω_3 ratios obtained by sequentially substituting each of the principal moments of inertia in equation 1.67. The principal axes are always axes of symmetry of the body. Equation 1.67 can be generalized as

$$\det(I_{\alpha\beta} - \lambda\delta_{\alpha\beta}) = 0 \qquad (1.68)$$

where $I_{\alpha\beta}$ is the inertia matrix, $\delta_{\alpha\beta}$ is the Kronecker delta and the λ_i are the eigenvalues or the principal moments of inertia of the body. The direction of each of the three principal axes ($i = 1, 2, 3$) is established by solving equation 1.67 for the corresponding angular velocity ratios $\omega_{i\alpha}/\omega_{i\beta}$ or equivalently, as in reference 8, by solving

$$(I_{\alpha\beta} - \lambda_i\delta_{\alpha\beta}) \begin{pmatrix} \cos \alpha_i \\ \cos \beta_i \\ \cos \gamma_i \end{pmatrix} = 0 \qquad (1.69)$$

for the direction cosines, which must also satisfy the condition

$$\cos^2 \alpha_i + \cos^2 \beta_i + \cos^2 \lambda_i = 1 . \qquad (1.70)$$

For example, consider the moments of inertia for a solid cube of side length "a" and mass density ρ with reference axes parallel to the sides of the cube as shown in Figure 1.11.

The moments of inertia are

$$I_{11} = \int_B (r_2^2 + r_3^2)\, dm$$
$$= \rho \int_o^a \int_o^a \int_o^a (r_2^2 + r_3^2)\, de_1\, de_2\, de_3$$
$$= \frac{2\rho a^5}{3}$$
$$= \frac{2Ma^2}{3} \qquad (1.71)$$

where the mass of the cube is $M = \rho a^3$; similarly, $I_{22} = I_{33} = I_{11}$ and

$$I_{12} = I_{21} = I_{31} = -\frac{Ma^2}{4} . \qquad (1.72)$$

To find the principal moments of inertia and principal axes, solve the characteristic determinant [2]

$$\begin{vmatrix} \alpha & \beta & \beta \\ \beta & \alpha & \beta \\ \beta & \beta & \alpha \end{vmatrix} = 0 \qquad (1.73)$$

where

$$\alpha = \left(\frac{2Ma^2}{3} - I \right)$$

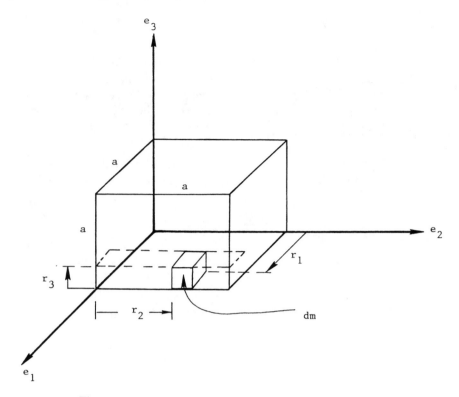

Figure 1.11   A cube of side length a and density $\rho = M/a^3$.

and

$$\beta = -\frac{Ma^2}{4}$$

Expanding the determinant in equation 1.73, the characteristic equation is

$$(\alpha - \beta)[\alpha(\alpha + \beta) - 2\beta^2] = 0 . \qquad (1.74)$$

One root of equation 1.74 is $\alpha = \beta$, yielding

$$I_1 = \frac{11Ma^2}{12} .$$

Also, the quadratic solution of equation 1.74

$$\alpha = \frac{-\beta \pm \sqrt{\beta^2 + 8\beta^2}}{2}$$

yields $\alpha = \beta$, from which one obtains

$$I_2 = \frac{11}{12} Ma^2 .$$

The other solution, $\alpha = -2\beta$ yields

$$I_3 = \frac{Ma^2}{6} .$$

If the origin of the reference axes is at the centroid of the cube, the reference axes are the principal axes due to symmetry. The corresponding principal moments of inertia about all axes passing through the centroid and oriented parallel to the sides of the cube are equal to $Ma^2/6$.

Substitution of the principal moments of inertia I_1, I_2, I_3 into Equation (1.67) yields the following ratios for the angular velocities

$$
\begin{aligned}
\omega_1 : \omega_2 : \omega_3 &= \quad\; 1 : -1 : 0 \quad (\text{for } I = I_1)\\
\omega_1 : \omega_2 : \omega_3 &= -1 : -1 : 2 \quad (\text{for } I = I_2)\\
\omega_1 : \omega_2 : \omega_3 &= \quad\; 1 : \quad 1 : 1 \quad (\text{for } I = I_3) .
\end{aligned}
$$

This shows that the third principal e_3' axis is the diagonal of the cube, and the other two axes are in a plane normal to e_3' with axis e_1' in the $e_1 e_2$ plane as is shown in Figure 1.12, for example.

When the cube is rotating about axis e_3', the angular momentum lies along this axis and has the value $h_3' = (1/6)Ma^2\omega$, where ω is the angular velocity. Similar results apply to the other principal axes of the cube.

1.3.3   Kinetic Energy

The kinetic energy, E_k, of a rigid body may be expressed as a dot product of angular momentum $\vec{h}$ and angular velocity $\vec{\omega}$ of the body [1]. Thus

$$
\begin{aligned}
E_k &= \frac{1}{2}\, \vec{\omega} \cdot \bar{\bar{I}} \cdot \vec{\omega}\\
&= \frac{1}{2}\, \vec{\omega} \cdot \vec{h}\\
&= \frac{1}{2}\, \omega' I \omega \qquad (1.75)
\end{aligned}
$$

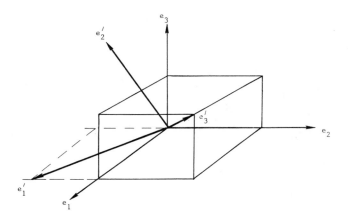

Figure 1.12 Principal axes for a cube.

where the transpose of a 3 × 1 column matrix is a 1 × 3 row matrix. That is,

$$\omega' = \begin{pmatrix} \omega_1 \\ \omega_2 \\ \omega_3 \end{pmatrix}' = (\omega_1, \omega_2, \omega_3) \qquad (1.76)$$

The body inertia matrix is given as

$$I = \begin{pmatrix} I_{11} & I_{12} & I_{13} \\ I_{21} & I_{22} & I_{23} \\ I_{31} & I_{32} & I_{33} \end{pmatrix} \qquad (1.77)$$

1.3.4 Euler's Dynamical Equations

Euler's dynamical equation is the equivalent of Newton's second law of motion for rotation about the center of mass. It is of the form

$$\frac{d\vec{h}}{dt} = \vec{T} \qquad (1.78)$$

where

$$\vec{h} = \bar{\bar{I}} \cdot \vec{\omega} \qquad (1.79)$$

is the angular momentum, and $\vec{T}$ is the sum of all external torques acting on the body.

The time derivative in equation 1.78 is with respect to an inertial reference frame. With respect to a body-fixed reference frame rotating with an angular velocity $\vec{\omega}$, Euler's equation becomes

$$\dot{\vec{h}} + \vec{\omega} \times \vec{h} = \vec{T}. \qquad (1.80)$$

The scalar component form of this equation can be expressed as

$$\dot{h}_\gamma + \omega_\alpha h_\beta \varepsilon_{\alpha\beta\gamma} = T_\gamma \qquad (1.81)$$

where α, β, γ vary from 1 to 3 and $\varepsilon_{\alpha\beta\gamma}$ is defined in equation 1.32. Expanded, equation 1.81 becomes

$$\dot{h}_1 + \omega_2 h_3 - \omega_3 h_2 = T_1$$
$$\dot{h}_2 + \omega_3 h_1 - \omega_1 h_3 = T_2$$
$$\dot{h}_3 + \omega_1 h_2 - \omega_2 h_1 = T_3 \qquad (1.82)$$

where h_1, h_2, h_3 are the angular momentum components along the body fixed coordinates $e_\alpha(\alpha = 1 - 3)$, and ω_1, ω_2, ω_3 are the angular velocity components about the body axes. $T_\alpha(\alpha = 1 - 3)$ are the body referenced external torque components.

For the case of the e_α axes being the principal axes of the body, the following simplification results

$$h_1 = I_1 \omega_1$$
$$h_2 = I_2 \omega_2$$
$$h_3 = I_3 \omega_3 \qquad (1.83)$$

where I_1, I_2, and I_3 are the corresponding principal moments of inertia of the body.

If the inertia dyadic is a function of time, then after differentiation, equation 1.80 can be written as

$$\bar{\bar{I}} \cdot \dot{\vec{\omega}} + \dot{\bar{\bar{I}}} \cdot \vec{\omega} + \vec{\omega} \times (\bar{\bar{I}} \cdot \vec{\omega}) = \vec{T}. \qquad (1.84)$$

In matrix notation this equation becomes

$$I\dot{\omega} + \dot{I}\omega + \bar{\omega} I\omega = T \qquad (1.85)$$

where I is the inertia matrix and

$$\bar{\omega} = \begin{pmatrix} 0 & -\omega_3 & \omega_2 \\ \omega_3 & 0 & -\omega_1 \\ -\omega_2 & \omega_1 & 0 \end{pmatrix} \qquad (1.86)$$

$$T = \begin{pmatrix} T_1 \\ T_2 \\ T_3 \end{pmatrix}. \qquad (1.87)$$

Equation 1.85 can be solved for the angular acceleration components $\dot{\omega}$ which can be integrated to obtain ω. The latter can then be used in the kinematical relations discussed previously to obtain either the Euler angles, the direction cosines, or the quaternions which specify the orientation (position) of the body frame relative to inertial space. A specification of position relative to a rotating frame can be an alternative desired result.

For example, if $\dot{I} = 0$, as in the case for a rigid body, then equation 1.85 can be solved to yield

$$\dot{\omega} = I^{-1}[T - \bar{\omega} I\omega] \qquad (1.88)$$

which can be integrated once to obtain ω. The results can be substituted into equation 1.46 which, in turn, can be integrated to obtain the quaternions q_i which describe the attitude of the rigid body as a function of time.

1.3.5 Torque-Free Body of Revolution Example

Consider a cylinder referenced to an inertial frame E_1, E_2, E_3 with principal axes e_1, e_2, e_3. The E_3 axis is placed along the system angular momentum vector $\vec{h}$, as shown in Figure 1.13, and the cylinder is given a spin of angular velocity $\vec{\omega}$. The values of the components of $\vec{\omega}$ along the e_1, e_2, e_3 axes are ω_1, ω_2, ω_3. The Euler angle rate values are $\dot{\psi}$, $\dot{\theta}$, and $\dot{\varphi}$. The values of the components of $\vec{h}$ along the e_3 axis and normal (transverse) to it are h_3 and h_t, respectively.

For a given initial angular velocity $\vec{\omega}$ at an angle γ relative to the e_3 axis, the motion of the body can be described in terms of the Euler angles and their rates.

Thus, for the case of zero external torques and dynamic symmetry, equation 1.82 can be written as

$$A\dot{\omega}_1 + (C - A)\omega_2\omega_3 = 0$$
$$A\dot{\omega}_2 + (A - C)\omega_1\omega_3 = 0$$
$$C\dot{\omega}_3 = 0 \qquad (1.89)$$

where

$A = I_1 = I_2$ is the centroidal principal moment of inertia about the transverse axis of the cylinder

and

$C = I_3$ is the principal moment of inertia about the longitudinal e_3 axis of the cylinder.

Differentiating the first equation in equation 1.89 and substituting $\dot{\omega}_2$ from the second yields

$$\ddot{\omega}_1 + \lambda^2\omega_1 = 0 \qquad (1.90)$$

From the third equation in equation 1.89, it is concluded that ω_3 must be a constant. Therefore substituting $\omega_3 = n$ gives

$$\lambda = \left(\frac{C - A}{A}\right)n .$$

The solution of equation 1.90 is of the form

$$\omega_1 = a \cos \lambda t + b \sin \lambda t \qquad (1.91)$$

where a, b are constants determined from initial conditions. Similarly, from the second equation in equation 1.89 the solution for ω_2 can be obtained in the form

$$\omega_2 = c \cos \lambda t + d \sin \lambda t \qquad (1.92)$$

where c and d are constants and λ is the frequency of the ω_1 and ω_2 rotations in the body fixed plane (e_1, e_2). Multiplying the first equation in equation 1.89 by ω_1, the second by ω_2, and adding yields

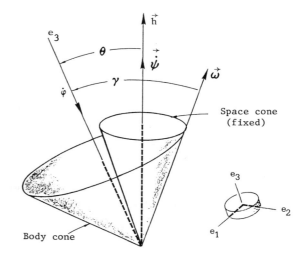

(a) Retrograde precession C > A.

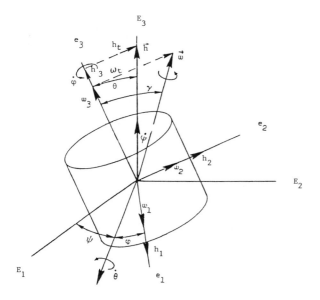

Figure 1.13 Torque-free motion for a body with dynamic or rotational symmetry (e.g., a cylinder).

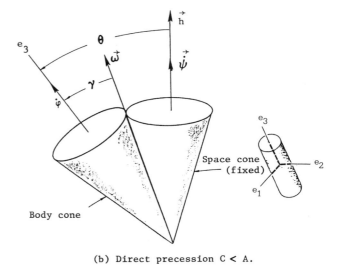

(b) Direct precession C < A.

Figure 1.14 Torque-free motion of an axis symmetrical rigid body.

$$\frac{d}{dt}(\omega_1^2 + \omega_2^2) = 0 \qquad (1.93)$$

which indicates that $\omega_1^2 + \omega_2^2$ is constant. Therefore we write the resultant angular velocity $\omega_t = $ constant where

$$\omega_t = (\omega_1^2 + \omega_2^2)^{1/2} . \qquad (1.94)$$

Consequently, the magnitude of the angular velocity is constant; i.e.,

$$\omega = (\omega_t^2 + n^2)^{1/2}$$
$$= \text{constant} . \qquad (1.95)$$

The nutation half-angle θ, and wobble half-angle γ are also constant since

$$\tan \theta = \frac{h_t}{h_3}$$
$$= \frac{A\omega_t}{C\omega_3}$$
$$= \text{constant} \qquad (1.96)$$

and

$$\tan \gamma = \frac{\omega_t}{\omega_3}$$
$$= \text{constant} . \qquad (1.97)$$

The results show that the body e_3 axis generates a body cone and a space cone by rotation of the axis e_3 about the angular momentum vector $\vec{h}$ which is fixed in inertial coordinates as illustrated in Figure 1.14. A body cone with a half-angle γ is also generated by the motion of the $\vec{\omega}$ vector about the e_3 axis. The rate of rotation of $\vec{\omega}$ in the e_α frame

is λ, which is also equal to the Euler angle rate $\dot{\varphi}$. The magnitude of the torque-free precession of the e_3 axis about $\vec{h}$ is

$$\dot{\psi} = \frac{C\dot{\varphi}}{(A - C)\cos\theta} \qquad (1.98)$$

as can be derived from the first equation in equation 1.89 [1].

1.4   References

1. Thomson, W. T. *Introduction to Space Dynamics*, Wiley, New York, 1961.
2. McCuskey, S. W. *Introduction to Advanced Dynamics*, Addison-Wesley, 1959.
3. Margulies, G. "On Real Four Parameter Representations of Satellite Attitude Motions," Philco/Ford Report No. 52, September 1963.
4. Kane, T. R., P. W. Likius, and D. A. Levinson. *Spacecraft Dynamics*, McGraw-Hill, 1983.
5. Hughes, P. C. *Spacecraft Attitude Dynamics*, John Wiley & Sons, 1986.
6. Mortensen, R. E. "On Systems for Automatic Control of the Rotation of a Rigid Body," Electronics Research Laboratory, University of California, Berkeley, Berkeley, CA, Report No. 63–23, November 1963.
7. Hankey, W. L., et al. "Use of Quaternions in Flight Mechanics," Air Force Wright Aeronautical Laboratories, TR-84–3045, March 1984.
8. Rimrott, F. P. J. *Introductory Attitude Dynamics*, Springer-Verlag, 1989.
9. Wertz, J. R., ed. *Spacecraft Attitude Determination and Control*, D. Reidel Publishing Co., 1980.
10. Grubin, C. "Derivation of the Quaternion Scheme via the Euler Axis and Angle," J. Spacecraft and Rockets, Vol. 7, No. 10, October 1970.
11. Brand, L. *Vector and Tensor Analysis*, Wiley, New York, 1947.

Chapter 2

Spin Stabilization

2.1 Introduction

Spin stabilization provides a simple means of obtaining a reasonably systematic attitude behavior of a spacecraft. Fairly simple instrumentation can be used to provide knowledge of attitude which widens the usefulness of such a spacecraft. Also, because of the comparatively large angular momentum contained in the spin, the spacecraft is highly resistant to external disturbing torques.

There are basically two types of spin-stabilized systems. The first type is simply a spinning body. The second type consists of two bodies rotating relative to each other. One body may consist of an Earth pointing antenna, while the other body provides a large counter rotating angular momentum.

Early communication satellites such as the Telstar, Syncom, and the Intelsats I and II were simple spinners. Intelsat IV, however, and a number of other spacecraft are of the dual-spin type. Spin stabilization is also widely used while coasting in parking and transfer orbits as well as during orbit injection burns. This chapter reviews the requirements of the attitude control system (ACS), response to external disturbances, and the methods used for spin control, nutation damping, and despin of the spin-stabilized spacecraft.

2.2 Control Requirements

The basic control requirements for typical spin-stabilized spacecraft are:

1. Deployment and spin-up
2. Sun/Earth acquisition
3. Spin and nutation control
4. Vehicle attitude (pointing) control.

For a vehicle with properly chosen moments of inertia, the control problem is primarily one of limiting vehicle nutation rates and pointing errors to the values established by the design criteria. Typical spin stabilization attitude control systems are shown schematically in Figure 2.1.

Initial deployment from a rocket stage is usually done by axial separation using springs. The spacecraft or the rocket stage is at that time spinning to minimize the effects of spring unbalances on the deployment. The spin-up can be accomplished by several spin-up jets or rockets. Sun and Earth acquisition is then accomplished by using the scanning motion of the body-mounted sensors. If a decrease in spin rate or a total despin of the spacecraft is required, this can be accomplished either by despin rockets or by the release of a tethered mass (yo-yo) which is cut off at the end of the maneuver and allowed to fly away from the spacecraft. The yo-yo method is an economical and effective approach which can reduce the spin speed of the spacecraft to any desired value.

Passive nutation control (damping) may be performed when the spin of the spacecraft is about its major principal axis. Any spin-up induced nutation can be effectively removed by energy dissipating dampers such as a ball-in-tube or a viscous fluid-type damper, for example. Active nutation control by jets or rockets may also be required if the spin is about the minor principal axis of the spacecraft. This condition may occur while in the transfer orbit. In the absence of active nutation control, the orientation of the spin axis due to energy dissipation may depart greatly from the desired direction and result in unacceptable performance during the orbit injection burn.

Vehicle attitude control (pointing) is achieved by periodic thrusting of the axial jet to precess the angular momentum of the spacecraft in the desired direction. Radial jets can be used for position control in the orbit.

2.3 Response of the Uncontrolled Vehicle

For the purpose of obtaining an insight into the performance of the uncontrolled vehicle, the spinning satellite can be approximated by a spinning cylinder.

The vehicle is spun up by spin rockets after separation from the launch vehicle. The spin-up propellant weight required is $W = I/gI_{sp}$; where I is the spin-up impulse, I_{sp} is the propellant specific impulse and g is the gravitational acceleration at sea level. The spin-up impulse is $I = F\Delta t$, where F is the spin-up thrust and Δt the duration of thruster firing. The firing duration Δt can be expressed as

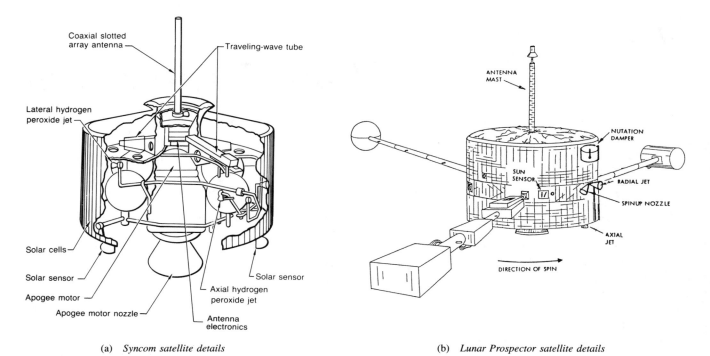

(a) *Syncom satellite details*   (b) *Lunar Prospector satellite details*

Figure 2.1   Spin stabilization systems.

$$\Delta t = \frac{C \, \Delta\omega}{FR} \qquad (2.1)$$

where

C = spin axis moment of inertia
R = thruster moment arm about the spin axis
$\Delta\omega$ = change in the spin angular speed in radians/second

Because of the imperfect operation of the spin-up control system, angular momentum will accumulate about the transverse axis to produce coning (or nutation) about the spin axis. To minimize coning, a nutation damper based on the principle of energy dissipation can be used. The dynamics can be examined with reference to Figure 2.2 as follows.

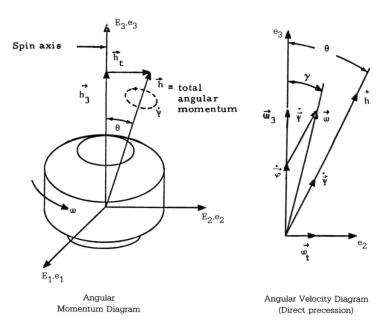

Angular
Momentum Diagram

Angular Velocity Diagram
(Direct precession)

Figure 2.2   Vector diagram illustrating the dynamics of a spinning cylinder.

Let the $\hat{e}_1$, $\hat{e}_2$, $\hat{e}_3$ unit vectors be directed along the principal axes of the cylinder which has moments of inertia, A, A, and C, respectively. For the case of the spin axis inertia $C > A$ and no external torques acting on the vehicle, the angular momentum, the angular velocity, and the spin axis, e_3, are all coincident. If now there is a component of angular momentum, h_t normal to the spin axis, the spin axis traces out a right circular cone with the axis of the cone coinciding with the new angular momentum vector (fixed in inertial space). This is referred to as torque-free precession or coning. The angle θ is called the nutation angle.

2.4 Precession of the Satellite

From the angular velocity diagram, it follows that

$$\vec{\omega} = \dot{\vec{\varphi}} + \dot{\vec{\psi}} \qquad (2.2)$$

where $\dot{\vec{\varphi}}$ is the Euler spin velocity along the e_3 body axis, and $\dot{\vec{\psi}}$ is the precession due to $\vec{h}_t$. The magnitudes of the vectors and their components are of the form

$$h_t = A\omega_t = A\omega \sin \gamma \qquad (2.3)$$
$$h_3 = C\omega \cos \gamma \qquad (2.4)$$
$$\tan \theta = \frac{h_t}{h_3} = \frac{A\omega \sin \gamma}{C\omega \cos \gamma} = \frac{A \tan \gamma}{C} \qquad (2.5)$$
$$h_3 = h \cos \theta = C(\dot{\varphi} + \dot{\psi} \cos \theta) = C\omega_3 . \qquad (2.6)$$

From this, it follows that the precessional velocity is

$$\dot{\psi} = \frac{\dot{\varphi}}{(A - C)\cos \theta} \qquad (2.7)$$

which gives

$$\dot{\varphi} = \frac{\dot{\psi}(A - C)}{C} \cos \theta . \qquad (2.8)$$

Therefore,

$$h_3 = A\dot{\psi} \cos \theta \qquad (2.9)$$

or

$$\dot{\psi} = \frac{h_3}{A \cos \theta} = \frac{H}{A} \approx \frac{h_3}{A} \qquad (2.10)$$

for $\theta \ll 1$. This is a useful small angle approximation for $\dot{\psi}$.

For $C > A$, retrograde precession occurs since $\dot{\psi}$ is negative, which causes $\gamma > \theta$. $\dot{\psi}$, however, is not a nutation rate. The nutation frequency λ is the angular velocity of the $\vec{\omega}$ vector as viewed by an observer stationed on the body at the axis of symmetry (e_3 axis). It is given by $\lambda = (C - A)\omega_3/A$.

If a body cannot be perfectly balanced, then the inertia matrix with respect to the mechanical spin axis will contain not only the diagonal terms $I_{11} = I_{22} = A$ and $I_{33} = C$, but also some off-diagonal or cross product terms $I_{\alpha\beta}$ for which

$\alpha \neq \beta$. The body will spin about its true principal axis which will, in general, be displaced from the mechanical axis by an angle δ, where

$$\delta \approx \frac{I_{\alpha\beta}}{C - A} . \qquad (2.11)$$

If, for example, an antenna is placed along such a mechanical axis on a spin stabilized satellite, then its pattern will oscillate through an angle equal to twice δ. This motion will be superimposed on that caused by spin-up errors ($\omega_t \neq 0$) and can be limited only by limiting the inertial cross product terms, $I_{\alpha\beta}$. Examples of torque free motion and its modeling are given in references 1–4.

2.5 Stability of Motion

From equation 1.75, the kinetic energy E_k of the body can be written as

$$E_k = \frac{1}{2} \vec{h} \cdot \vec{\omega}$$
$$= \frac{1}{2} (h_3 \hat{e}_3 + h_t \hat{e}_2) \cdot (\omega_3 \hat{e}_3 + \omega_t \hat{e}_2)$$
$$= \frac{1}{2} (h_3 \omega_3 + h_t \omega_t)$$
$$= \frac{1}{2} (C\omega_3^2 + A\omega_t^2)$$
$$= \frac{1}{2} h\omega \cos(\theta - \gamma)$$
$$= \frac{1}{2} h\omega(\cos \theta \cos \gamma + \sin \theta \sin \gamma) \qquad (2.12)$$

where

$$\sin \gamma = \frac{\omega_t}{\omega}, \quad \cos \gamma = \frac{h_3}{C\omega}, \quad \text{and } \cos \theta = \frac{h_3}{h} .$$

Therefore,

$$E_k = \frac{h^2}{2C} \left[1 + \left(\frac{C - A}{A} \right) \sin^2 \theta \right] . \qquad (2.13)$$

Differentiating with respect to time yields the rate of energy dissipation (in joules/sec or watts)

$$\dot{E}_k = \frac{dE_k}{dt} = \frac{h^2}{C} \left(\frac{C - A}{A} \right) \sin \theta \cos \theta \dot{\theta} . \qquad (2.14)$$

For a rate of energy dissipation $dE_k/dt < 0$, $\dot{\theta} < 0$ if $C - A > 0$. Thus, the angle θ decreases until the minimum energy state is reached. Conversely, if $C - A < 0$ the angle θ increases until the body spins about its transverse axis (flat spin) which again corresponds to the state of minimum kinetic energy.

2.6 Passive Nutation Control

For bodies spinning about their major axes, the nutation half-angle θ can be reduced to zero by an energy-dissipating device (damper). One type of damper is a ball-in-tube system in which the ball moves in a tube in response to the nutation accelerations, and the energy dissipation takes place (either by friction or viscous effects of a gas within the tube). The damping time constant τ can be obtained for such as system using the following energy sink approach.

Consider a viscous damper of the ball in a tube type as shown in Figure 2.3.

The differential work done by the ball in a viscous fluid on the satellite is

$$dW = -k\dot{x}\,dx \qquad (2.15)$$

where the magnitude of the viscous damping force is

$$f = k\dot{x}. \qquad (2.16)$$

Here

k = damping coefficient
dx = differential displacement of the ball from equilibrium
$\dot{x}$ = instantaneous velocity (magnitude only) of the ball.

Writing equation 2.15 as

$$dW = -k\dot{x}\frac{dx}{dt}\,dt$$
$$= -k\dot{x}^2\,dt \qquad (2.17)$$

the average work per cycle can be obtained by integrating; namely,

$$W_{av} = -\frac{k}{P}\int_0^P \dot{x}^2\,dt \qquad (2.18)$$

where P is the period of damper motion.

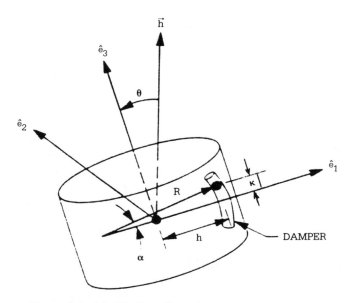

Figure 2.3 Model of satellite using a ball-in-tube damper.

The average work per cycle is equal to the rate of change of kinetic energy, i.e., $W_{av} = \dot{E}_k$. Therefore, from equation 2.14

$$W_{av} = \frac{h^2}{C}\left(\frac{C-A}{A}\right)\dot{\theta}\sin\theta\cos\theta = \dot{E}_k \qquad (2.14)$$

$$\approx \frac{h^2}{C}\left(\frac{C-A}{A}\right)\theta\dot{\theta} \qquad \text{for} \qquad \theta << 1. \qquad (2.19)$$

To solve equation 2.18 and obtain θ as a function of time, an explicit relation for $\dot{x}$ in equation 2.18 must be obtained. This can be done by solving for $\dot{x}$ under varying simplifying assumptions. In general, however, after the integration in equation 2.18 is performed, W_{av} is often expressed in the form

$$W_{av} = -\theta^2 f(\alpha_1, \alpha_2 \ldots \alpha_n) \qquad (2.20)$$

where $f(a_1, a_2 \ldots a_n)$ represents a function of the system parameters. In that case equation 2.19 becomes

$$\frac{h^2}{C}\left(\frac{C-A}{A}\right)\theta\dot{\theta} + \theta^2 f = 0 \qquad (2.21)$$

or

$$\dot{\theta} + \frac{f\theta}{\dfrac{h^2}{C}\left(\dfrac{C}{A}-1\right)} = 0. \qquad (2.22)$$

This is of the form

$$\dot{\theta} + \frac{\theta}{\tau} = 0 \qquad (2.23)$$

where the system damping time constant τ is

$$\tau = \frac{h^2}{fC}\left(\frac{C}{A}-1\right). \qquad (2.24)$$

The solution of equation 2.23 is of the form

$$\theta = \theta_0 e^{-t/\tau} \qquad (2.25)$$

where

$$\theta_o = \text{initial nutation angle.}$$

For example, at $\tau = t$ seconds

$$\frac{\theta}{\theta_0} = \frac{1}{e}. \qquad (2.26)$$

If, by energy dissipation within the satellite (without external torques), the ω_t term is eliminated, the final spin speed of the satellite will be greater because angular momentum is conserved. Thus, if the initial angular momentum h is given by

$$h = (h_3^2 + h_t^2)^{1/2}$$
$$h = [(C\omega_3)^2 + (A\omega_t)^2]^{1/2}, \qquad (2.27)$$

then it must be equal to $C\omega_f$, where ω_f is the final spin speed after removal of ω_t.

Thus,

$$(C\omega_f)^2 = (C\omega_3)^2 + (A\omega_t)^2$$

$$\omega_f = [\omega_3^2 + \left(\frac{A}{C}\right)^2 \omega_t^2]^{1/2}$$

$$> \omega_3 \qquad\qquad (2.28)$$

which shows that the final spin axis angular velocity has increased due to nutation damping.

2.7 Active Nutation Damping

Upper stage rockets can spin about an axis of minimum moment of inertia in transfer orbit. In the presence of energy dissipation, the symmetry axis will eventually depart from the angular momentum vector. If the energy dissipation is sufficiently low, as is the case for some satellites, the divergence of the symmetry axis from the angular momentum vector can be quite small, resulting in acceptable performance. Conversely, if the amount of energy dissipation is

sufficiently large, then there can be a substantial deviation of the symmetry axis from the angular momentum vector. This can result in unacceptably large injection errors.

This is illustrated in Figure 2.4. The rate at which the body departs from pure spin motion about an axis of minimum moment of inertia depends upon the inertial properties of the body under consideration, the spin speed, and the rate at which kinetic energy is being dissipated.

If the rate at which kinetic energy is being dissipated is sufficiently slow, then the spinning spacecraft might ostensibly remain in a state of nearly pure spin with only minor departure from that state for long periods of time. Conversely, highly dissipative spacecraft that are nominally in a state of pure spin about an axis of minimum moment of inertia can be expected to depart from that state more rapidly by comparison. A measure that is used to characterize the amount of departure is the nutation angle θ.

Accurate predictions of nutation divergence behavior require that all major energy dissipation sources within the

$$I_1 > I_2 > I_3$$

$$2T = I_1\omega_1^2 + I_2\omega_2^2 + I_3\omega_3^2$$

$$h^2 = I_1^2\omega_1^2 + I_2^2\omega_2^2 + I_3^2\omega_3^2$$

MAXIMUM
KINETIC ENERGY STATE
(pure spin motion)

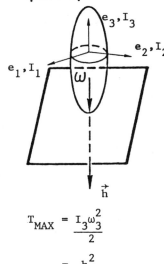

$$T_{MAX} = \frac{I_3\omega_3^2}{2}$$

$$= \frac{h^2}{2I_3}$$

INTERMEDIATE
KINETIC ENERGY STATE

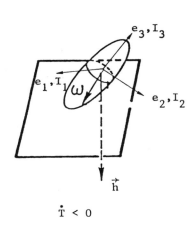

$$\dot{T} < 0$$

MINIMUM
KINETIC ENERGY STATE
(FLAT SPIN)

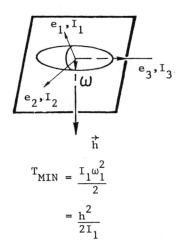

$$T_{MIN} = \frac{I_1\omega_1^2}{2}$$

$$= \frac{h^2}{2I_1}$$

Figure 2.4 Effects of energy dissipation on the motion of a prolate spheroid.

spacecraft be correctly identified and correctly characterized. Active nutation damping can be performed to reduce
or eliminate the nutation angle θ when the spin is about the
minor axis of the spacecraft in the presence of energy dissipation.

The methodology of actively controlling nutation is as
follows: First, a sensor detects the presence of nutation and
establishes the phase and amplitude of the motion with
respect to a rotor-fixed coordinate system. Next, the sensor
output signal is threshold detected, amplified, and converted
to a jet command. Finally, the axial jet fires once per nutation cycle for some fraction of the cycle which results in
the application of a transverse torque in opposition to the
nutational motion.

The basic technique used for active nutation control is
illustrated in Figure 2.5.

As the transverse rate vector, $\vec{\omega}_t$, rotates clockwise at rotor
nutation frequency λ with respect to a rotor-fixed (e_1, e_2)
plane, a rotor-fixed jet develops a torque impulse which,
on the average, opposes the sense of the $\vec{\omega}_t$ vector; thus,
the residual transverse angular momentum is removed and
the nutation angle decreases. A system for active nutation
is shown in Figure 2.6.

2.8 Vehicle Reorientation in Space

If it is required to reorient a spinning body in space for a
proper orbit injection attitude or to correct its orientation,
then an axial thruster (parallel to the spacecraft spin vector)
can be used to produce a torque about an axis normal to
the spin vector.

Consider the precession geometry in Figure 2.7.

The governing Euler's equation for torque is

$$\dot{\vec{h}} + \dot{\vec{\psi}} \times \vec{h} = \vec{T} \tag{2.29}$$

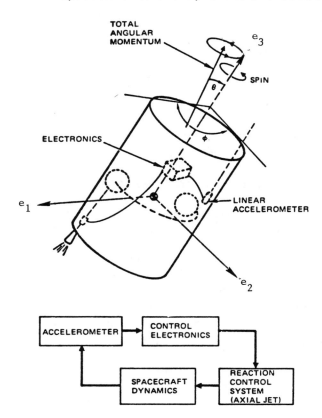

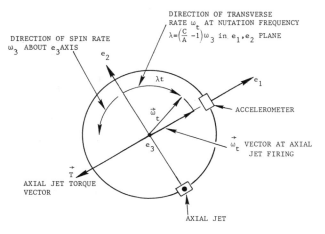

Figure 2.5 Active nutation damping.

Figure 2.6 An active nutation control system.

where $\dot{\vec{\psi}}$ is the rotation (precession) of the body frame of
reference e_α with respect to the fixed reference frame E_β,
(α, $\beta = 1 - 3$).

An average gyroscopic precession $\dot{\psi}_{av}$ about the E_1 axis
will result if an average external torque T_{av} is applied about
the E_2 axis. This can be done by firing an axial thruster
parallel to the spin axis e_3 over a firing angle $\Delta\phi$, where
$\pm\Delta\phi/2$ is measured from the E_1 direction.

For a constant body spin rate ω_3 the magnitude of the
angular momentum is $h = C\omega_3$. Therefore,

$$\dot{\psi}_{av} = \frac{T_{av}}{h}$$
$$\approx \frac{\Delta\psi}{\Delta t} . \tag{2.30}$$

The precession angle per firing interval Δt is

$$\Delta\psi \approx \frac{T_{av}\Delta t}{h} . \tag{2.31}$$

Here

$$T_{av} = \frac{2FR}{\Delta\phi} \int_{o}^{\Delta\phi/2} (\cos \phi) \, d\phi = \frac{2FR}{\Delta\phi} \sin\left(\frac{\Delta\phi}{2}\right) \tag{2.32}$$

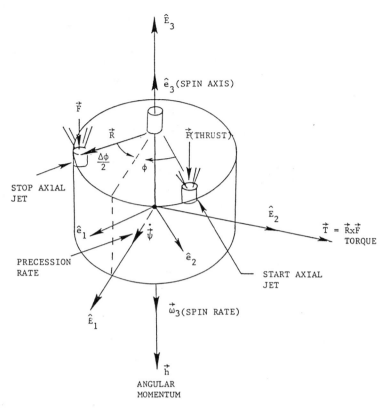

Figure 2.7 Pulsed-jet control for spin stabilized spacecraft.

and

$$\Delta t = \frac{\Delta \phi}{\omega_3} . \qquad (2.33)$$

The amount of propellant required to produce the total angular rotation $\psi = n\Delta\psi$, where n is the number of thruster firings, is

$$W = nF\Delta t/gI_{sp} . \qquad (2.34)$$

I_{sp} is the specific impulse of the propellant, and g is the gravitational acceleration at sea level. For additional discussions on vehicle reorientation in space, see references 1–3.

2.9 Spin-Rate Decrease Due to Fuel Expenditure (Jet Damping)

Consider a spinning satellite as shown in Figure 2.8.

The angular momentum of the satellite and the remaining fuel about the spin axis at any time t is

$$H(t) = I(t)\,\omega(t) \qquad (2.35)$$

where the total moment of inertia (satellite plus propellant) is given as

$$I(t) = I_{sat} + mr_{cm}^2 . \qquad (2.36)$$

Here the propellant mass m at any time t is related to its initial mass m_o according to $m = m_o - \dot{m}t$

Each particle of fuel that leaves the spinning satellite adds to the angular momentum $h(t)$ of the fuel that has already left the satellite. Thus, this becomes

$$h(t_o) = \Delta mr_e^2\omega_o$$
$$h(t_1) = h(t_o) + \Delta mr_e^2\omega_1$$

at time t_1, or

$$h(t_n) = \Delta mr_e^2\omega_o + \Delta mr_e^2\omega_1 \ldots \qquad (2.37)$$
$$+ \Delta mr_e^2\omega_n$$

at time t_n.

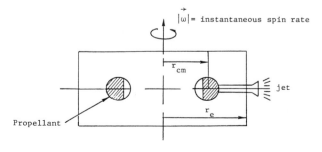

Figure 2.8 Jet damping schematic diagram.

Here

Δm = mass of escaping fuel
r_e = radius at which fuel escapes
ω_i = satellite angular velocity at time t_i.

Forming the incremental change in h during a time increment Δt and passing to the limit yields

$$\dot{h}(t) = \dot{m} r_e^2 \omega \qquad (2.38)$$

where $\dot{m}$ is assumed constant. If no external torques act on the satellite, the system angular momentum is conserved. Thus

$$\dot{H}(t) + \dot{h}(t) = 0 . \qquad (2.39)$$

That is

$$I\dot{\omega} + \dot{I}\omega + \dot{m} r_e^2 \omega = 0 \qquad (2.40)$$

where

$$I = I_s + (m_o - \dot{m}t) r_{cm}^2 . \qquad (2.41)$$

2.9.1 Approximate Solution

Ignoring equation 2.40 but using conservation of angular momentum gives

$$(I_s + m_o r_{cm}^2)\omega_o = (I_s + m_o r_e^2)\omega_e \qquad (2.42)$$

where ω_o is the initial spin rate and ω_e is the final spin rate.
Solving for ω_e yields

$$\omega_e = \left(\frac{I_s + m_o r_{cm}^2}{I_s + m_o r_e^2} \right) \omega_o$$

$$= \left(1 + \frac{m_o r_{cm}^2}{I_s} \right)\left(1 - \frac{m_o r_e^2}{I_s} + \cdots \right) \omega_o$$

or

$$\frac{\omega_e - \omega_o}{\omega_o} = -\frac{(m_o r_e^2 - m_o r_{cm}^2)}{I_s} \qquad (2.43)$$

$$= \frac{m_o}{I_s}(r_{cm}^2 - r_e^2) < 0 .$$

Thus, with an initial angular rate of ω_o, the final angular rate ω_e is less than ω_o. This solution essentially assumes that the entire fuel mass moves instantaneously and as a whole.

The solution to equation 2.40, on the other hand is of the form

$$\omega = \frac{I_o \omega_o}{I} \exp\left(-r_e^2 \int_{m_o}^{m} \frac{dm}{I} \right)$$

where the zero subscript denotes initial condition. It can be shown, for example, that for r_{cm} and r_e constant the effect of the rotating jet is that of a damper. The resulting solution is similar to that of equation 2.43. For radially burning solid rockets, however, ω may increase under certain conditions (see references 3 and 9). An example of the effect of jet damping on the precessional motion of a spinning rocket can be found in reference 10.

2.10 Yo-yo Despin

The problem of despinning a spacecraft or reducing its spin to a specified value by releasing yo-yo-like weights on a cord can be separated into two phases. In Phase 1 the cord extends tangentially relative to the spacecraft. In Phase 2 the cord length is constant as its position changes from tangent to radial with respect to the spinning spacecraft as is shown in Figure 2.9. Releasing two symmetrically oriented cords after Phase 2 is more efficient weight-wise than releasing after Phase 1 [3, 5].

Assuming conservation of angular momentum and kinetic energy, the governing equations are, respectively,

$$\vec{h} = I_s \vec{\omega} + \vec{r} \times m\vec{v}$$
$$= \vec{h}_o$$

and

$$E_k = \frac{1}{2} I_s \omega^2 + \frac{1}{2} mv^2$$
$$= E_{ko} \qquad (2.44)$$

where $\vec{h}_o$ is the initial angular momentum vector and E_{ko} is the initial kinetic energy.

Here

I_s = moment of inertia of the body about the spin axis
m = total mass of the two weights
ω, v = instantaneous angular and linear velocities of the body and weights, respectively.

The solution to equation 2.44 (from reference 1) yields the spin rate of the body at any time t in Phase 1 as

$$\omega = \omega_o \left(\frac{K - \omega_o^2 t^2}{K + \omega_o^2 t^2} \right) . \qquad (2.45)$$

For a spacecraft of radius R, ω_o is the initial spin rate and

$$K = \frac{I_s}{mR^2} + 1 . \qquad (2.46)$$

The length of wire (assumed to be massless) as a function of time (as given in reference 5) is

$$\ell = R\omega_o t . \qquad (2.47)$$

Also, the maximum tension during Phase 1 is

$$F_{max} = 1.3\, m\omega_o^2 \lambda(1 - R^2/\lambda^2) \qquad (2.48)$$

where

$$\lambda^2 = \frac{I_s}{m} + R^2$$
$$= KR^2 . \qquad (2.49)$$

The length of wire for terminal angular velocity ω_f during Phase 1 (tangential deployment) is

$$\ell_f = R \sqrt{K\frac{(\omega_o - \omega_f)}{(\omega_o + \omega_f)}} . \qquad (2.50)$$

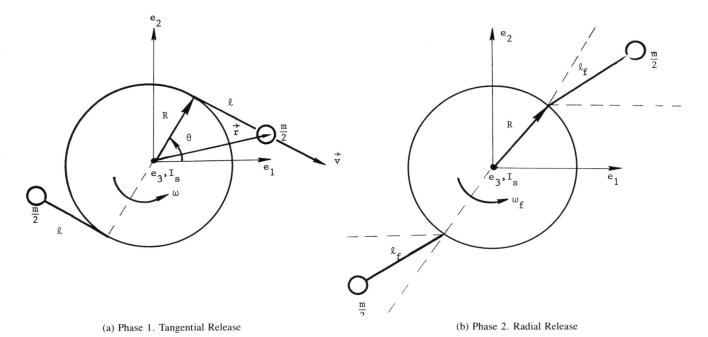

(a) Phase 1. Tangential Release   (b) Phase 2. Radial Release

Figure 2.9   Yo-yo despin mechanism.

For $\omega_f = 0$, this becomes

$$\ell_f = R\sqrt{K} . \qquad (2.51)$$

For radial release (Phase 2) the terminal length for complete despin is given as [1]

$$\ell_f = R(\sqrt{K} - 1) . \qquad (2.52)$$

An approximate solution for radial release including the effect of the mass of the cord is [5]

$$\frac{1 + (\omega_f/\omega_o)}{1 - (\omega_f/\omega_o)} = \frac{I_s}{m(R + \ell_f)^2} \qquad (2.53)$$

where

ℓ_f = final length of the cord
m = total mass of both weights plus
        1/3 the mass of the cords.

2.11 Vehicle Response to External Disturbances

Consider, for example, a spinning body with principal moments of inertia A, B, and C which is subject to external torques T_1, T_2, and T_3 acting along the principal axes e_α with unit vectors $\hat{e}_\alpha$. Euler's dynamical equation which was presented earlier in equation 1.78 is reintroduced here as

$$\frac{d\vec{h}}{dt} = \vec{T} \qquad (2.54)$$

or, with respect to the body axes, e_α

$$\dot{\vec{h}} + \vec{\omega} \times \vec{h} = \vec{T} . \qquad (2.55)$$

Here $\vec{h}$ is the angular momentum, and

$$\vec{T} = T_1\hat{e}_1 + T_2\hat{e}_2 + T_3\hat{e}_3$$

and

$$\vec{\omega} = \omega_1\hat{e}_1 + \omega_2\hat{e}_2 + \omega_3\hat{e}_3 .$$

The components of equation 2.55 can be expressed in scalar form as in equation 1.82. Thus

$$\dot{h}_1 + \omega_2 h_3 - \omega_3 h_2 = T_1$$
$$\dot{h}_2 + \omega_3 h_1 - \omega_1 h_3 = T_2$$
$$\dot{h}_3 + \omega_1 h_2 - \omega_2 h_1 = T_3 \qquad (2.56)$$

where

$$h_1 = A\omega_1$$
$$h_2 = B\omega_2$$
$$h_3 = C\omega_3 .$$

For a dynamically symmetrical body, $A = B$. Therefore, equation 2.56 becomes

$$A\dot{\omega}_1 + (C - A)\omega_2\omega_3 = T_1$$
$$A\dot{\omega}_2 + (A - C)\omega_1\omega_3 = T_2$$
$$C\dot{\omega}_3 = T_3 \qquad (2.57)$$

For the case of zero external torque about the e_3 axis (that is, $T_3 = 0$ and $\omega_3 = $ constant), equation 2.57 becomes

$$\dot{\omega}_1 + \lambda\omega_2 = T_1/A$$
$$\dot{\omega}_2 - \lambda\omega_1 = T_2/A \qquad (2.58)$$

where

$$\lambda = (C - A)\omega_3/A$$
$$= \text{torque-free nutation frequency} \quad (2.59)$$
[the rate at which vector $\vec{\omega}$
rotates in the (e_1, e_2) body plane]

2.11.1 Response to a Torque Impulse

A torque impulse $T\Delta t$ applied about the e_1 axis (with $T_2 = T_3 = 0$) is equivalent to the application of the following initial angular velocities: namely

$$\omega_1(0) = T_1\Delta t/A$$
$$\omega_2(0) = 0 \quad (2.60)$$

Solving equation 2.58 with the initial conditions of equation 2.60 results in

$$\omega_1 = \omega_1(0)\cos \lambda t$$
$$\omega_2 = \omega_1(0)\sin \lambda t \ . \quad (2.61)$$

This shows that the amplitude of the body transverse angular velocity ω_t is

$$\omega_t = (\omega^2_1 + \omega^2_2)^{1/2}$$
$$= \omega_1(0) \ . \quad (2.62)$$

In view of equation 2.61 and 2.62, it can be seen that ω_t can be expressed as

$$\omega_t = \omega_1(0)e^{-i\lambda t} \ . \quad (2.63)$$

Thus, ω_t rotates in the body plane with angular velocity λ.

2.11.2 Response to Body-Fixed Torques

Consider, for example, a dynamically symmetrical body ($A = B$) spinning about its principal axis e_3 which is precessing about a mean reference direction E_3 in inertial space. At time $t = 0$ the ignition of a motor causes small transverse torques due to a thrust line offset from the mass center of the body along the e_1 and e_2 principal axes. It is desired to examine the angular velocity of the body and the motion of the spin vector relative to inertial space. The geometry of the body is illustrated in Figure 2.10.

2.11.2.1 Equations of Motion

The equations of motion (equation 2.57) for the system in Figure 2.10 are

$$\dot{\omega}_1 + (\mu - 1)\omega_2\omega_3 = T_1/A$$
$$\dot{\omega}_2 + (1 - \mu)\omega_1\omega_3 = T_2/A$$
$$\dot{\omega}_3 = 0 \quad (2.64)$$

where $\mu = C/A$.

From this it follows that

$$\omega_3 = n$$
$$= \text{constant} \ . \quad (2.65)$$

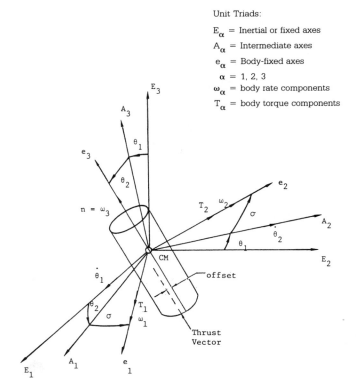

Unit Triads:

E_α = Inertial or fixed axes
A_α = Intermediate axes
e_α = Body-fixed axes
α = 1, 2, 3
ω_α = body rate components
T_α = body torque components

Figure 2.10   Attitude angles for offset thrust.

Therefore, equation 2.64 can be written as

$$\dot{\omega}_1 + \lambda\omega_2 = T_1/A$$
$$\dot{\omega}_2 - \lambda\omega_1 = T_2/A \quad (2.66)$$

where, as before,

$$\lambda = (\mu - 1)\omega_3 \ .$$

Letting

$$\omega_t = \omega_1 - i\omega_2$$
$$T_t = T_1 - iT_2 \quad (2.67)$$

where $i = \sqrt{-1}$. Equation 2.66 can be written as

$$\dot{\omega}_t + i\lambda\omega_t = T_t/A \ . \quad (2.68)$$

The solution of this differential equation for ω_t is

$$\omega_t = \omega_t(0)e^{-i\lambda t} + \frac{T_t}{i\lambda A}(1 - e^{-i\lambda t}) \quad (2.69)$$

where $\omega_t(0)$ is the initial value of ω_t.

2.11.3 Kinematical Relationships

Referring to Figure 2.10 it can be seen that for small offset angles θ_1 and θ_2 of the spin axis e_3 relative to the fixed reference direction $\hat{E}_3$, the body referenced angular rate components ω_α, ($\alpha = 1 - 3$) are approximately related to the respective angular velocity components $\dot{\theta}_\beta(\beta = 1, 2)$ through the rotation matrix

$$\begin{pmatrix} \omega_1 \\ \omega_2 \\ \omega_3 \end{pmatrix} \approx \begin{pmatrix} \cos\sigma & \sin\sigma & 0 \\ -\sin\sigma & \cos\sigma & 0 \\ 0 & 0 & 1 \end{pmatrix} \begin{pmatrix} \dot{\theta}_1 \\ \dot{\theta}_2 \\ n \end{pmatrix} \qquad (2.70)$$

In terms of the offset angular velocities, this yields

$$\dot{\theta}_1 \approx \omega_1 \cos\sigma - \omega_2 \sin\sigma \qquad (2.71)$$
$$\dot{\theta}_2 \approx \omega_1 \sin\sigma + \omega_2 \cos\sigma$$

where $\sigma = nt$ is the angle about the spin axis and θ_1 and θ_2 are treated as "small" angles.

Substituting

$$\theta = \theta_1 - i\theta_2 \qquad (2.72)$$

into equation 2.71 gives

$$\dot{\theta} = \omega_t(\cos\sigma - i\sin\sigma)$$
$$= \omega_t e^{-i\sigma} . \qquad (2.73)$$

2.11.4 Sequence of Motions

Before motor ignition $(t < 0)$ there are no external torques $(T_t = 0)$ and the solution for the motion of the spin axis can be obtained as follows: Substituting equation 2.69 into equation 2.73 and recalling that $\sigma = nt$ yields

$$\dot{\theta} = \omega_t(0)e^{-i\lambda t} \cdot e^{-int} \qquad (2.74)$$
$$= \omega_t(0)e^{-i\mu nt}$$

where

$$\lambda = (\mu - 1)n \text{ as before} .$$

Now integrating equation 2.74 gives us

$$\theta = \frac{i\omega_t(0)e^{-i\mu nt}}{\mu n}$$
$$= \eta_o e^{-i\mu nt} \qquad (2.75)$$

where

$$\eta_o = \frac{i\omega_t(0)}{\mu n}$$
$$= \text{initial nutation half angle} .$$

The body is thus seen to nutate at a constant (complex) nutation half-angle η_o at frequency μn.

At time $t = 0$, a step transverse constant torque T_t is applied. In view of equation 2.68, equation 2.73 becomes

$$\dot{\theta} = \omega_t e^{-int}$$
$$= \left[\omega_t(0)e^{-i\lambda t} + \frac{T_t}{i\lambda A}(1 - e^{-i\lambda t}) \right] e^{-int}$$
$$= \omega_t(0)e^{-i\mu nt} + \frac{T_t}{i\lambda A}(e^{-int} - e^{-i\mu nt}) \qquad (2.76)$$

which, when integrated, yields

$$\theta = \eta_o e^{-i\mu nt} - \eta^* \left[1 + \frac{\mu}{1 - \mu}e^{-int} - \frac{e^{-i\mu nt}}{1 - \mu} \right] \qquad (2.77)$$

where

$$\eta^* = \frac{T_t}{n^2 C} . \qquad (2.78)$$

The body motion described by equations 2.69 and 2.77 can be expressed in the form

$$\omega_t = \omega_A e^{-i(\lambda t - \xi_D)} + \omega_B e^{i\xi_E} \qquad (2.79)$$
$$\theta = \theta_A e^{-i(\mu nt - \xi_A)} + \theta_B e^{-i(nt - \xi_B)} + \theta_C e^{i\xi_C} \qquad (2.80)$$

where

$$\omega_A = \mu n\theta_A \qquad (2.81)$$
$$\omega_B = n\theta_B . \qquad (2.82)$$

also,

$$\theta_A = \left| \eta_o + \frac{\eta^*}{1 - \mu} \right| \qquad (2.83)$$
$$= \text{final nutation amplitude at frequency } \mu n,$$

$$\theta_B = \left| -\frac{\mu\eta^*}{1 - \mu} \right| \qquad (2.84)$$
$$= \text{final "wobble" amplitude at frequency } n,$$

and

$$\theta_C = \left| -\eta^* \right| \qquad (2.85)$$
$$= \text{offset angle of the spin axis} .$$

The phase angles ξ_A, ξ_B, ξ_C, ξ_D and ξ_E are generally arbitrary and are dependent on vehicle design, thrust offset, and the initial conditions. For example, an instantaneous relationship defining the quantity θ_A is illustrated in Figure 2.11.

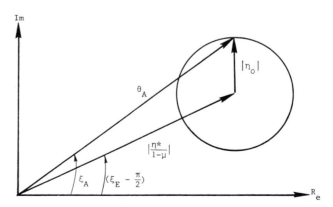

Figure 2.11 Instantaneous relationship between final nutation amplitude θ_A, phase angle ξ_A, and $|\eta_o|$.

Since the torque can be applied at any time in the nutation cycle, the phase angle of η_o in Figure 2.11 has a uniform probability distribution over 0 to 2π. If ξ_E is also assumed to be uniformly distributed over the range 0 to 2π, then the median value for θ_A (where the probability distribution of θ_A is equal to 0.5) is of the form

$$\overline{\theta}_A = \sqrt{|\eta_o|^2 + \left|\frac{\eta^*}{1-\mu}\right|^2} \qquad (2.86)$$

The value of θ_A will, however, vary between the limits of

$$\left||\eta_o| - \left|\frac{\eta^*}{1-\mu}\right|\right| \le \theta_A \le |\eta_o| + \left|\frac{\eta^*}{1-\mu}\right| \qquad (2.87)$$

If, for simplicity, it is assumed that all phase angles are zero, then the equations 2.79 and 2.80 can be written in the form

$$\omega_t \approx \omega_A e^{-i\lambda t} + \omega_B \qquad (2.88)$$
$$\theta \approx \theta_A e^{-i\mu nt} + \theta_B e^{-int} + \theta_C . \qquad (2.89)$$

The physical interpretation of equations 2.88 and 2.89 is as follows:

1. The application of step function torque T_t at $t = 0$ results in a constant transverse angular velocity ω_B.
2. The amplitude of the body angular velocity at nutation frequency is instantly changed from $\mu n \eta_o$ to ω_A.

3. The motion of the spin axis in inertial space is changed: the nutation amplitude is changed from η_o to θ_A. A new "wobble" of amplitude θ_B appears at frequency n and an offset θ_C of the spin axis is produced. For typical thrust offset angles θ_B is usually small ($< 1°$).

Since $\theta = \theta_1 - i\theta_2$, the real and imaginary parts of θ in equation 2.89 yield the normalized spin axis angles

$$\phi_1 = \frac{\theta_1}{\theta_B} = \frac{\theta_A}{\theta_B} \cos \mu nt + \cos nt + \frac{\theta_C}{\theta_B} \qquad (2.90)$$

$$\phi_2 = \frac{\theta_2}{\theta_B} = \frac{\theta_A}{\theta_B} \sin \mu nt + \sin nt \qquad (2.91)$$

where

$$\frac{\theta_A}{\theta_B} = \left\{\left[\frac{|\eta_o|(C-A)n^2}{|T_t|}\right]^2 + \left(\frac{A}{C}\right)^2\right\}^{1/2} . \qquad (2.92)$$

For the case of zero initial nutation angle ($\eta_o = 0$)

$$\phi_1 = \frac{A}{C} \cos\left(\frac{Cnt}{A}\right) + \cos nt + \frac{A-C}{C} \qquad (2.93)$$

$$\phi_2 = \frac{A}{C} \sin\left(\frac{Cnt}{A}\right) + \sin nt . \qquad (2.94)$$

Equations 2.93 and 2.94 are plotted in Figures 2.12 and 2.13 for $A/C = 15$ and 0.5 corresponding to a rod and a disk-like body, respectively.

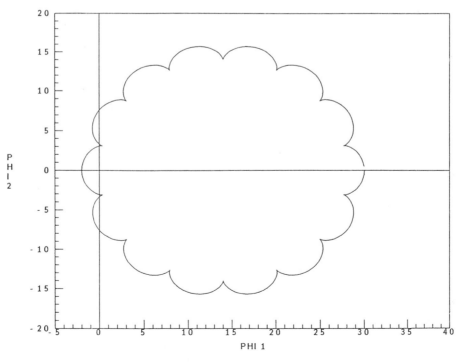

Figure 2.12 Plot of equations 2.93 and 2.94 for a rod ($A/C = 15$).

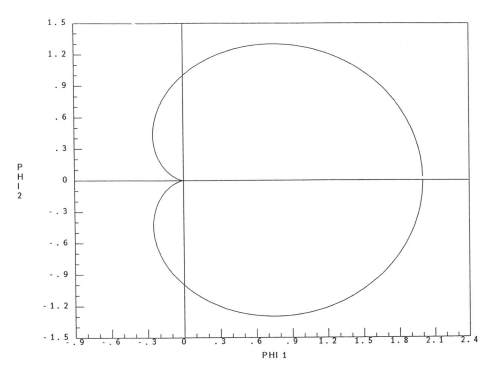

Figure 2.13   Plot of equations 2.93 and 2.94 for a disk ($A/C = 0.5$).

2.12   Separation Analysis

2.12.1   Ideal Separation for Spinning Bodies

Consider two symmetrical spinning bodies separating under the influences of only an axial velocity impulse. If a spring provides the force of separation, assume that no angular rates are induced by the spring during the process.

Before separation, the motion is a nutation about a given reference axis. The connected bodies nutate with a half cone angle η_o about the total angular momentum vector, $\vec{h}$, of the system. The nutation angle as a function of time can be expressed as

$$\eta = \eta_o e^{i\mu n t} \qquad (2.95)$$

where

$$\eta_o = \frac{\omega_t}{\mu n} = \frac{\omega_t A}{Cn}$$

and

ω_t = transverse angular rate
n = spin rate about the axis of symmetry
C = spin moment of inertia
A = transverse moment of inertia
$\mu = C/A$.

The geometry before separation is illustrated in Figure 2.14a.

In Figure 2.14, $\vec{h}_s = Cn\hat{e}_3$, $\vec{h}_t = A\vec{\omega}_t$, $\vec{V}_1 = -d_1\vec{\omega}_t$, and $\vec{V}_2 = d_2\vec{\omega}_t$. The $\vec{V}_1$ and $\vec{V}_2$ instantaneous lateral velocity components are the result of nutation of the combined configuration of the two bodies. After separation, the bodies move axially at the applied rate (e.g., from compressed springs). Also, because of the presence of the lateral velocities the bodies move in opposite directions along the E_2 direction as shown in Figure 2.14b.

The angular velocities of the bodies are the same before and after separation. Conservation of angular momentum requires that

$$\vec{h} = \vec{h}_s + \vec{h}_t \qquad (2.96)$$
$$\vec{h}_1 = \vec{h}_{1s} + \vec{h}_{1t} \qquad (2.97)$$
$$\vec{h}_2 = \vec{h}_{2s} + \vec{h}_{2t} \qquad (2.98)$$

and

$$\vec{h} = \vec{h}_1 + \vec{h}_2 + \vec{h}_t \qquad (2.99)$$

where $\vec{h}_t$ is the lateral angular momentum arising from the separating bodies' transverse velocity components. The angular velocity and angular momentum vectors are illustrated in Figures 2.15(a) and 2.15(b), respectively.

The preceding relationship can be proven as follows:

The composite spin moment of inertia is

$$C = C_1 + C_2 \qquad (2.100)$$

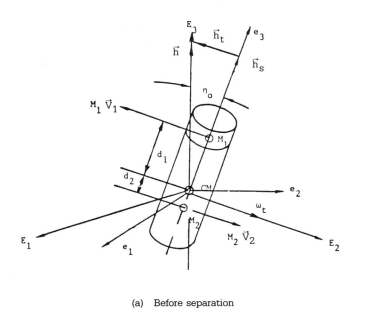

(a) Before separation   (b) After separation

Figure 2.14 Geometry before and after separation of two spinning bodies.

and the composite transverse moment of inertia is

$$A = A_1 + M_1 d_1^2 + A_2 + M_2 d_2^2 . \quad (2.101)$$

Therefore,

$$\begin{aligned}
\vec{h} &= C\vec{n} + A\vec{\omega}_t \\
&= (C_1 + C_2)\vec{n} + (A_1 + A_2 + M_1 d_1^2 \\
&\quad + M_2 d_2^2)\vec{\omega}_t \\
&= \vec{h}_1 + \vec{h}_2 + \vec{h}_r
\end{aligned} \quad (2.102)$$

where

$$\begin{aligned}
\vec{h}_1 &= C_1\vec{n} + A_1\vec{\omega}_t \\
\vec{h}_2 &= C_2\vec{n} + A_2\vec{\omega}_t \\
\vec{h}_r &= (M_1 d_1^2 + M_2 d_2^2)\vec{\omega}_t .
\end{aligned} \quad (2.103)$$

In conclusion, the nutation angles of bodies 1 and 2 after separation compared to the nutation angle η_o of the combined system are given by the expressions

$$\tan \eta_o = \frac{h_t}{h_s} = \frac{A\omega_t}{Cn} \quad (2.104)$$

$$\tan \eta_1 = \frac{h_{1t}}{h_{1s}} = \frac{A_1\omega_t}{C_1 n} \quad (2.105)$$

$$\tan \eta_2 = \frac{h_{2t}}{h_{2s}} = \frac{A_2\omega_t}{C_2 n} \quad (2.106)$$

therefore,

$$\frac{\tan \eta_1}{\tan \eta_o} \approx \frac{\eta_1}{\eta_o} = \frac{A_1 C}{C_1 A} \quad (2.107)$$

$$\frac{\tan \eta_2}{\tan \eta_o} \approx \frac{\eta_2}{\eta_o} = \frac{A_2 C}{C_2 A} . \quad (2.108)$$

2.13 Recommended Practice

Design and actual flight experience has resulted in the following practices which should be considered to ensure satisfactory attitude control of spin stabilized spacecraft [6–8].

2.13.1 General Practices

1. Maintain current and accurate properties of the spacecraft and alternate configurations.
2. Determine the mass and balance of the spacecraft.
3. Define all sources of thrust misalignment including:
   a. Mechanical misalignment of the thruster.
   b. Compliance of the thruster support structure.
4. Provide adequate gyroscopic "stiffness" (spin stability maintainability) through a sufficiently high spin rate or sufficiently large spin-axis moment of inertia to prevent significant disturbance of the angular momentum vector. If the spacecraft mass properties change during the mission, care should be exercised to preclude spin-to-tranverse axis moment-of-inertia ratios approaching values near 1.0. Specifically, an approximate rule of thumb for all spin controlled spacecraft is to maintain the relationship

$$1.05 < \frac{I_{spin}}{I_{transverse}} < 0.95$$

with due regard to the stabilization requirements (no energy dissipation) for inertia ratios less than 1.0.
5. Keep track of inertia ratios and the location of the center of mass.

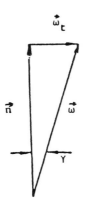

(A) Angular velocity resolution

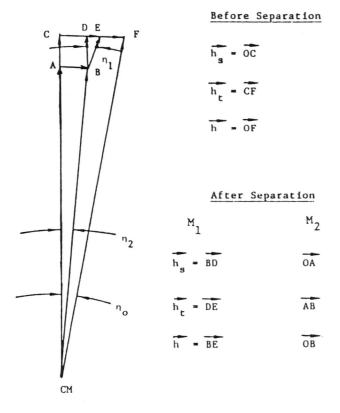

Before Separation

$$\vec{h}_s = \overrightarrow{OC}$$

$$\vec{h}_t = \overrightarrow{CF}$$

$$\vec{h} = \overrightarrow{OF}$$

After Separation

$$M_1 \qquad\qquad M_2$$

$$\vec{h}_s = \overrightarrow{BD} \qquad \overrightarrow{OA}$$

$$\vec{h}_t = \overrightarrow{DE} \qquad \overrightarrow{AB}$$

$$\vec{h} = \overrightarrow{BE} \qquad \overrightarrow{OB}$$

(B) Angular momentum resolution

Figure 2.15 Angular velocity and angular momentum resolution.

6. Consider all effects of thrust on spin dynamics, including nutation, spin-axis orientation, spin rate, and attitude perturbation.

7. Allow sufficient firing time to permit the thrust to reach its steady-state value.

8. Determine the effects of energy dissipation on the spacecraft stability.

9. Apply thrust by either of two commonly used techniques:

   a. To use axial thrust such as for apogee boost, first orient the spin axis (nominally identical with the thrust axis) to the desired velocity increment direction. A continuous axial thrust is applied until the desired velocity change is achieved. The spacecraft is then reoriented to the desired attitude. This technique should not be used when propellant expenditure for reorientation is a significant fraction of the propellant expended during the thrusting maneuver.

   b. To apply small velocity increments such as for midcourse correction, leave the spacecraft with its spin axis in its nominal orientation. Thrust is applied continuously in the axial direction and pulsed in the radial direction, so that the vector sum of the component velocity increments is the desired velocity increment.

Account for engine thrust effects on spin rate. If spin rate is critical, bias intial spin rate to account for thrust effects.

2.13.2 Compact Near-Rigid Body

1. Whenever possible, spin-stabilize about the axis of maximum moment of inertia.

2. If the spacecraft is expected to maintain its spin about an axis of minimum moment of inertia for an extended period of time, then all possible sources of energy dissipation are to be avoided. The following are specific recommended practices:

   a. The spacecraft should be as rigid as possible.

   b. If there is flexibility in the spacecraft body or appendages, care should be taken that the resulting motion does not give rise to excessive mechanical energy dissipation.

   c. Fluids (such as propellants) that could dissipate energy in a sloshing mode should be avoided.

   d. Passive nutation dampers should be caged.

3. If the spacecraft is expected to maintain its spin about an axis of minimum moment of inertia for an extended period and energy dissipation cannot be prevented, then an appropriate source of energy should be incorporated. Active nutation dampers are representative of the type of devices to be utilized.

4. If the spacecraft is expected to maintain its spin about an axis of maximum moment of inertia, then it should be designed so that the interaction of the flexible appendages with the space environment decreases the ratio of angular momentum to angular kinetic energy.

2.13.3 Spin Resonance

To avoid spin resonances, which are always undesirable, the following considerations should be followed:

1. Spin rates should not exceed 70 percent of the lowest natural frequency of the transverse bending mode. This margin reflects uncertainties in the values of the natural frequency and of the spin rate which may be achieved.

2. Spin rates above the natural frequencies of the lower modes are not recommended. If, however, a spin rate above the lower natural frequencies is required, then:
   a. The rate should be nearly midway between adjacent natural frequencies if practical.
   b. The rate should be separated from the nearest natural frequency by a margin equal to at least 30 percent of the lowest natural frequency.
   c. The spin acceleration should be high enough that the bending deformation developed during passage of the spin rate past the natural frequencies will not be excessive.

3. Bending deformation should be minimized by an arrangement of vehicle frame and internal components that will result in local centers of mass being on the centerline of the undeformed vehicle along the entire length.

2.14   References

1. Kaplan, M. H. *Modern Spacecraft Dynamics and Control*, John Wiley & Sons, 1976.

2. Wertz, J. R., ed. *Spacecraft Attitude Determination and Control*, D. Reidel Publishing Co., 1980.

3. Thomson, W. T. *Introduction to Space Dynamics*, John Wiley & Sons, 1961.

4. Hughes, P. C. *Spacecraft Attitude Dynamics*, John Wiley & Sons, 1986.

5. Fedor, J. V. "Theory and Design Curves for a Yo-Yo De-Spin Mechanism for Satellites," NASA TN D-708, August 1961.

6. "Spacecraft Attitude Control During Thrusting Maneuvers," NASA SP-8059, February 1971.

7. "Effects of Structural Flexibility on Spacecraft Control Systems," NASA SP-8016, April 1969.

8. Agrawal, B. N. *Design of Geosynchronous Spacecraft*, Prentice-Hall, Inc., 1986.

9. Breuer, D. W., and Southerland, W. R. "Jet-Damping Effects: Theory and Experiment," Journal of Spacecraft and Rockets, Vol. 2, No. 4, July–August 1965, pp. 638–639.

10. Thomson, W. T., and Reiter, G. S. "Jet Damping of a Solid Rocket: Theory and Flight Results," AIAA Journal, Vol. 3, No. 3, March 1965, pp. 413–417.

Chapter 3
Dual-Spin Stabilization

3.1 Introduction

Dual-spin stabilization of attitude control is often used when the ability to point instruments or provide a platform with good spin stabilization is required. Stabilization is possible for large length to diameter ratios about the axis of minimum moment of inertia. The dual-spin satellite is a "gyrostat," i.e., a body with constant inertial configuration which may include some internal angular momenta.

Problems may arise due to structural flexibility, unbalancing effects, or bearing assembly friction between the platform and the rotor. A schematic diagram of a dual-spin concept is shown in Figure 3.1, and two examples of spacecraft utilizing dual-spin stabilization are illustrated in Figure 3.2.

3.2 Design Considerations for a Dual-Spin Spacecraft

For a dual-spin stabilized spacecraft the following functions must generally be provided:

1. A means to impart spin to the spinning section without inducing excessive nutation or attitude errors
2. A means to remove spacecraft nutation within a reasonable time period
3. A means to determine spin axis orientation
4. A means to command and achieve spin axis reorientation or velocity control

The functional operation of the system is as follows. After separation from the booster, rotor spin-up, and platform despin, the nutation damper (possibly consisting of some viscous fluid in a tube) removes all induced nutational motion. Such motion can be induced by separation, apogee engine firing, or other maneuvers. The outputs of on-board attitude and position sensors are processed by a computer for corrective control and/or required velocity corrections. These commands are then transmitted to the jet control electronics, which properly phase the reaction control system (RCS) jet firings with respect to the spacecraft spin speed and its orientation to the Earth.

3.3 Sensing Subsystem

The attitude control sensing subsystem functions are (1) to provide attitude measurement and spin data for computation of the roll/yaw attitude error and momentum variation, and (2) to provide real-time horizon-crossing data as an input to the pitch-axis stabilization control subsystem.

Such a sensing system might consist of three infrared (IR) Earth-horizon detectors mounted on the spinner. Two of these detectors would form a symmetric "VEE" configuration about the local vertical, while the third sensor would be mounted perpendicular to the spin axis to provide inputs to the pitch axis control subsystem.

When the horizon-to-horizon pulses from the two "VEE" sensors are of equal duration, then the spin axis is parallel to the surface of the Earth. If a yaw error (rotation about the local vertical) exists at this time, then the effect will be evidenced as a roll error of the same magnitude, but occurring 90 degrees later in the orbit because of the inertial stability of the momentum axis. The yaw and roll errors can thus be determined and corrected by a reaction control system.

3.4 Despin Control System

The despun platform attitude control system (ACS) consists of two major subsystems. The first subsystem provides control about the roll and yaw axes, while the second controls the position about the pitch axis. This control can be effected through a closed-loop system comprising a simple infrared (IR) sensor, compensation networks, and a torque motor which exchanges momentum between a flywheel mounted on an axis parallel with the spacecraft's pitch axis and the main spacecraft structure. A third subsystem provides control of the spacecraft pitch axis momentum by reaction jets.

The pitch axis of the spacecraft is nominally directed along the orbit normal. The yaw axis is always along the outward vertical, and the roll axis is along the orbital ve-

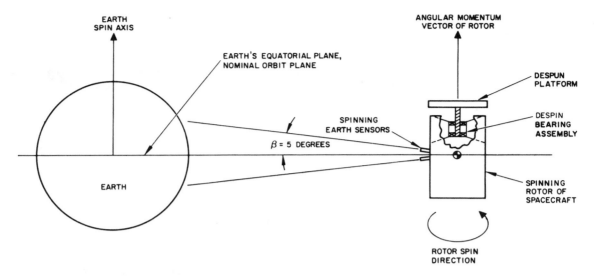

Figure 3.1 Dual-spin concept.

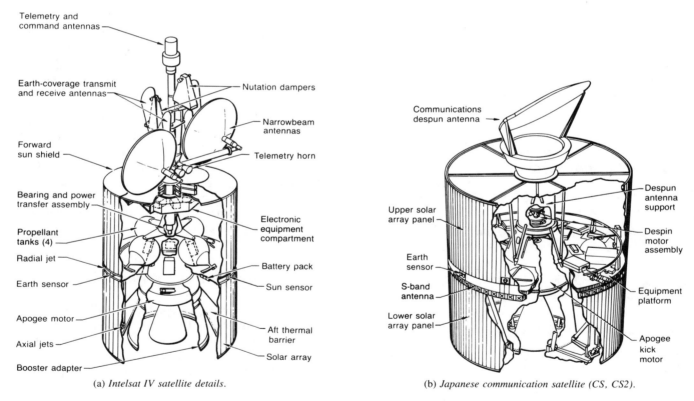

(a) *Intelsat IV satellite details.*   (b) *Japanese communication satellite (CS, CS2).*

Figure 3.2 Dual-spin stabilization examples.

locity vector. The stabilization pointing accuracy depends upon the sensor accuracy, satellite dynamic balancing, and the impulse size of the mass expulsion (gas jet) subsystem. An overall functional block diagram for a satellite ACS is shown in Figure 3.3.

The pitch axis position control (despin servo) system can be a standard sampled data type. This type of subsystem is required to despin the payload relative to the spinner, to search and capture the desired line of sight (LOS) direction, and to accurately maintain this direction for a long period

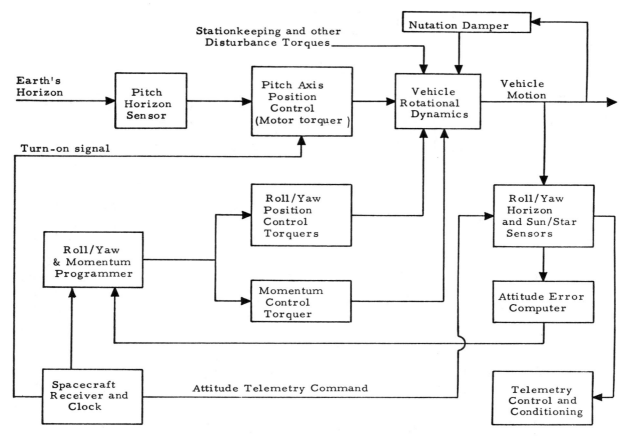

Figure 3.3 Functional block diagram of a satellite's attitude control system (ACS).

of time. System reliability and minimum power consumption require a minimum complexity and a motor employing a minimum torque capability compatible with steady-state running loads and system stability margins. High pointing accuracy requires that all biases and input disturbances be minimized.

An optimum design would be a hybrid configuration which operates as a single loop during the steady-state track mode and one which uses a secondary (rate) loop only for LOS acquisition. Such a control system is shown in Figure 3.4. In this design the electrical power is derived from solar cells mounted on the despun section of the spacecraft. For larger power requirements or more efficient generation, Sun-oriented panels (or other power sources) may be utilized.

The rotating assembly consists of an outer housing, a pair of ball bearings, a servo motor, a shaft, and a slipring-brush unit. A labyrinth seal arrangement can also be used. A number of sliprings can keep the current flow per unit area at low levels.

3.5 Momentum and Reaction Jet Sizing

The spin rate of the satellite is a free variable affected by the following factors:

1. Desired resistance to external torques (attitude correction frequency)
2. Required containment of motor thrust misalignment and axial jet offset
3. Desired stiffness during possible payload slewing maneuvers
4. Wearout life of bearings, brushes, and sliprings
5. Energy (hence, fuel mass) required to create angular momentum

A spin rate of 10 to 60 rpm is consistent with most factors above and yields a wearout life for the rotating assembly measured in years.

Thrust magnitudes and propellant mass, required for velocity and attitude control, are determined chiefly by the attitude correction resolution and the total momentum

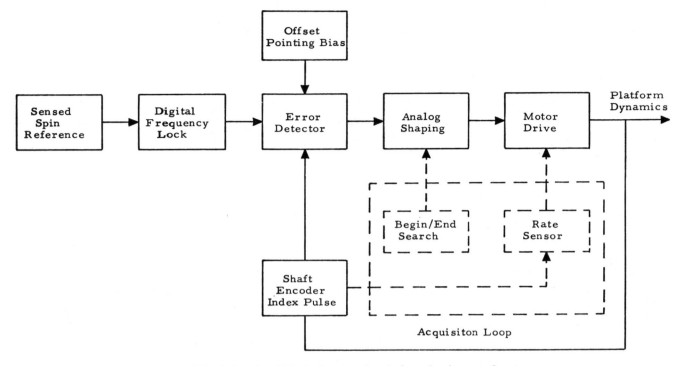

Figure 3.4 A functional block diagram of a platform despin control system.

change requirements, respectively. A precision system might have large axial and radial jet thrusters for velocity control and a small additional axial jet for precession torquing.

3.6 Equations of Motion and Equilibrium

Equilibrium with respect to a reference frame occurs when all the noncyclic internal variables are equal to zero and when some body-fixed frame $e_\alpha (\alpha = 1, 2, 3)$ has the same angular velocity as an orbiting reference frame E_α. In terms of total angular momentum $\vec{h}$ and external torque $\vec{T}$, the equation of motion for the satellite relative to inertial space is (from equation 1.78)

$$\frac{d\vec{h}}{dt} = \vec{T} = \text{sum of all external torques} \qquad (3.1)$$

where

$\vec{h} = \bar{\bar{I}} \cdot \vec{\omega} + \vec{h}_{in}$
$\bar{\bar{I}}$ = inertia dyadic of the satellite
$\vec{\omega}$ = angular velocity of the satellite (e_α reference frame)
$\vec{h}_{in}$ = internal angular momentum .

Expanded, equation 3.1 becomes

$$\dot{\bar{\bar{I}}} \cdot \vec{\omega} + \bar{\bar{I}} \cdot \dot{\vec{\omega}} + \vec{\omega} x (\bar{\bar{I}} \cdot \vec{\omega})$$
$$+ \dot{\vec{h}}_{in} + \vec{\omega} \times \vec{h}_{in} = \vec{T} \qquad (3.2)$$

where the time differentiation is with respect to the body fixed axes e_α rotating at the angular velocity $\vec{\omega}$ measured relative to an inertial reference frame. In matrix notation, equation 3.1 can be written as

$$\dot{h} + \tilde{\omega}h = T \qquad (3.3)$$

where

$$h = \begin{pmatrix} h_1 \\ h_2 \\ h_3 \end{pmatrix} = \begin{array}{l} \text{body components of total} \\ \text{angular momentum} \end{array}$$

$$\tilde{\omega} = \begin{pmatrix} 0 & -\omega_3 & \omega_2 \\ \omega_3 & 0 & -\omega_1 \\ -\omega_2 & \omega_1 & 0 \end{pmatrix}$$

and

$$T = \begin{pmatrix} T_1 \\ T_2 \\ T_3 \end{pmatrix} = \text{body components of external torques .}$$

In equation 3.2, the inertia dyadic $\bar{\bar{I}}$ can be written as

$$\bar{\bar{I}} = \begin{pmatrix} I_{11}\hat{e}_1\hat{e}_1 & I_{12}\hat{e}_1\hat{e}_2 & I_{13}\hat{e}_1\hat{e}_3 \\ I_{21}\hat{e}_2\hat{e}_1 & I_{22}\hat{e}_2\hat{e}_2 & I_{23}\hat{e}_2\hat{e}_3 \\ I_{31}\hat{e}_3\hat{e}_1 & I_{32}\hat{e}_3\hat{e}_2 & I_{33}\hat{e}_3\hat{e}_3 \end{pmatrix}$$

$$= I_{\alpha\beta}\hat{e}_\alpha\hat{e}_\beta$$

where $\hat{e}_{\alpha\beta}(\alpha, \beta = 1 - 3)$ are the dyads of the unit vectors and $I_{\alpha\beta}$ are the components of the satellite inertia matrix. A diagram of the system (gyrostat) is shown in Figure 3.5.

3.7 Simplified Dual-Spin System Dynamics

The following analysis presents the simplest description of the dual-spin system dynamics to illustrate the basic characteristics of its motion. A derivation of the despun part nutation frequency λ_1 and the nutation angle θ will be presented. Dual-spin system stability criteria will then be outlined based on the energy-sink method of analysis.

3.7.1 Perfectly Symmetric System

Consider an idealized dual-spin satellite as shown in Figure 3.6.

Let the components of the system angular momentum along the $e_\alpha(\alpha = 1 - 3)$ axes (fixed in the platform) be

$$h_1 = A\omega_1$$
$$h_2 = A\omega_2 \qquad (3.4)$$
$$h_3 = C_1\omega_3 + C_2\Omega$$

where

$$A = \text{total transverse moment of inertia}$$
$$C_1, C_2 = \text{platform and spinner inertias along}$$
$$\text{the } e_3 \text{ axis, respectively}$$
$$\omega_\alpha(\alpha = 1 - 3) = \text{inertial angular velocities of}$$
$$\text{the platform}$$
$$\Omega = \text{inertial angular velocity of rotor}.$$

The Euler equations (see equation 1.82) for torque free motion in scalar form are

$$\dot{h}_1 + \omega_2 h_3 - \omega_3 h_2 = 0$$
$$\dot{h}_2 + \omega_3 h_1 - \omega_1 h_3 = 0 \qquad (3.5)$$
$$\dot{h}_3 + \omega_1 h_2 - \omega_2 h_1 = 0.$$

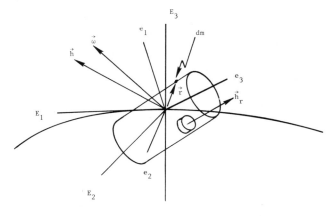

Figure 3.5 Gyrostat diagram.

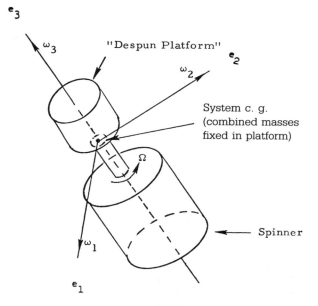

Figure 3.6 Dual-spin configuration.

Substituting equation 3.4 into equation 3.5, assuming $\dot{\Omega} = 0$ (constant spin), and linearizing the resultant equations yields

$$\dot{\omega}_1 + \lambda_1\omega_2 = 0$$
$$\dot{\omega}_2 - \lambda_1\omega_1 = 0 \qquad (3.6)$$

where

$$\lambda_1 = \frac{C_2\Omega + C_1\omega_3}{A} - \omega_3$$

and

$$C_1\dot{\omega}_3 = 0.$$

This last result indicates that ω_3 is constant.

The solution of equation 3.6 yields

$$\omega_1 = \omega_1(0) \cos \lambda_1 t$$
$$\omega_2 = \omega_2(0) \sin \lambda_1 t \qquad (3.7)$$

where $\omega_1(0)$ and $\omega_2(0)$ are the initial values of ω_1 and ω_2 respectively. Here λ_1 represents the angular velocity of the transverse rate ω_t as seen in the platform (inertial nutation rate). Thus, the amplitude of ω_t is

$$\omega_t = (\omega_1^2 + \omega_2^2)^{1/2}$$

This result is shown in Figure 3.7 where the angle θ is the nutation angle and $\vec{h}$ is the fixed angular momentum vector whose amplitude is given by

$$h = (h_1^2 + h_2^2 + h_3^2)^{1/2}$$
$$= \sqrt{(A\omega_t)^2 + (C_1\omega_3 + C_2\Omega)^2}. \qquad (3.8)$$

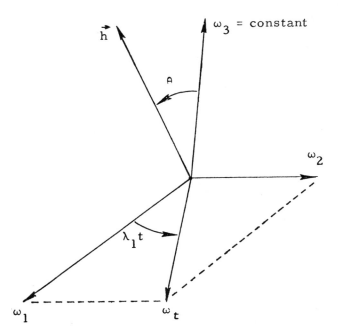

Figure 3.7 Nutational geometry.

Consequently,

$$\tan \theta = \frac{A\omega_t}{C_1\omega_3 + C_2\Omega}. \qquad (3.9)$$

For a despun platform, $\omega_3 \equiv 0$.

3.8 Dual-Spin System Stability Considerations

The spinning body stability analysis discussed previously showed that in the presence of energy dissipation the spin is stable if it occurs about the axis of maximum moment of inertia and unstable if it occurs about the axis of minimum moment of inertia. To satisfy launch constraints, solar power requirements, or other constraints, the question of spin (nutational) stability is of primary importance. As will now be shown, an ideal dual-spin satellite can be stabilized if sufficiente damping is provided on the despun part of the spacecraft (platform) to offset any destabilizing damping effects on the spinning part. The following analysis will be based on the energy-sink method [1–4].

Consider again the idealized model shown in Figures 3.6 and 3.7. The nutation angle θ for the case of nonzero transverse angular velocity ω_t is given by equation 3.9 and can be determined directly by finding the changes in the transverse and spin speeds when the energy is dissipated within the system. The energy dissipation can occur because of structural flexing (damping), vibration, or fuel damping mechanisms on either the spinning or despun parts of the satellite. Frictional losses in the shaft coupling the two bodies should not be included since a servo loop controlled motor is usually employed to counteract these effects.

For a perfectly rigid system, the kinetic energy E_k is conserved (constant) and can be written as

$$2E_k = A\omega_t^2 + C_1\omega_3^2 + C_2\Omega^2$$
$$= \text{Constant} . \qquad (3.10)$$

Now, with energy dissipation, kinetic energy is not constant but is a negative function of time. Thus

$$\frac{dE_k}{dt} = A\omega_t\dot{\omega}_t + C_1\omega_3\dot{\omega}_3 + I\Omega\dot{\Omega} < 0$$
$$= \dot{E}_{k1} + \dot{E}_{k2} \qquad (3.11)$$

where $\dot{E}_{k1}$ and $\dot{E}_{k2}$ are the rates of energy dissipation for bodies 1 and 2, respectively.

For zero external torques, the angular momentum is constant, therefore,

$$\dot{\vec{h}} = 0 ,$$

and equation 3.8 yields

$$\frac{d}{dt}[(A\omega_t)^2 + (C_1\omega_3 + C_2\Omega)^2] = 0$$

or

$$A^2\omega_t\dot{\omega}_t + (C_1\omega_3 + C_2\Omega)(C_1\dot{\omega}_3 + C_2\dot{\Omega}) = 0 . \qquad (3.12)$$

Equations 3.11 and 3.12 must be satisfied simultaneously. These equations can be combined to form

$$\dot{E}_{k1} + \dot{E}_{k2} = -\lambda_1 C_1\dot{\omega}_3 - \lambda_2 C_2\dot{\Omega} \qquad (3.13)$$

where

$$\lambda_1 = \frac{C_1\omega_3 + C_2\Omega}{A} - \omega_3$$
$$\lambda_2 = \frac{C_1\omega_3 + C_2\Omega}{A} - \Omega .$$

Recognizing that equation 3.11 is a sum of two equations; namely,

$$C_1\dot{\omega}_3 = -\frac{\dot{E}_{k1}}{\lambda_1}$$
$$C_2\dot{\Omega} = -\frac{\dot{E}_{k2}}{\lambda_2}$$

which when substituded into equation 3.12 results in

$$A\dot{\omega}_t\omega_t = \lambda_o\left(\frac{\dot{E}_{k1}}{\lambda_1} + \frac{\dot{E}_{k2}}{\lambda_2}\right)$$
$$= \frac{d}{dt}\frac{A\omega_t^2}{2} = \text{kinetic energy of nutation} \qquad (3.14)$$

where

$$\lambda_o = \frac{C_1\omega_3 + C_2\Omega}{A}$$
$$= \frac{h_3}{A} > 0$$
$$= \text{total system nutation frequency (precession) .}$$

Equation 3.14 shows that $\omega_t < 0$ (θ decreases) when

$$\frac{\dot{E}_{k1}}{\lambda_1} + \frac{\dot{E}_{k2}}{\lambda_2} < 0 \qquad (3.15)$$

which is a useful stability criterion.

For a satellite with an essentially despun platform, ω_3 is equal to the orbital rate or is negligibly small so that the excitation frequency on the platform is

$$\lambda_1 \approx \frac{C_2 \Omega}{A} \qquad (3.16)$$

and the excitation frequency on the rotor is

$$\lambda_2 \approx \frac{C_2 \Omega}{A} - \Omega$$

$$= \left(\frac{C_2}{A} - 1 \right) \Omega . \qquad (3.17)$$

The stability criterion now becomes

$$\frac{\dot{E}_{k1}}{\frac{C_2}{A}} + \frac{\dot{E}_{k2}}{\left(\frac{C_2}{A} - 1 \right)} < 0 . \qquad (3.18)$$

Therefore, for a satellite with spin moment of inertia $C_2 > A$ (which represents a favorable inertia distribution) the stability criterion is always satisfied (for any $\dot{E}_{k1}$ or $\dot{E}_{k2}$ which are negative by definition). This means that a damper can be placed on either body to damp out nutation.

For a satellite with an unfavorable inertia distribution ($C_2/A < 1$), the quantity $\dot{E}_{k2}/(C_2/A - 1)$ is positive if $\dot{E}_{k2}$ is negative (energy dissipation on the rotor) so that $\dot{E}_{k1}/(C_2/A)$ must be a larger negative quantity for stability. This can be achieved by placing a large energy dissipator (damper) on the despun part ($\dot{E}_{k1} < 0$) to overcome the destabilizing damper on the spinning part.

3.9 Practical Implications

The experience with dual-spin satellites to date indicates the importance of taking into account all sources of energy dissipation on a spacecraft if nutational stability is to be achieved. The energy-sink method can be used to determine the required amount of damping on the despun part of the satellite when fuel sloshing, structural flexibilities, and the bearing energy dissipation are evaluated. If insufficient damping is provided on the despun part (platform), then an unstable nutational motion (θ increasing) can result. A limit cycle on the order of a degree or more may be reached ($\theta = $ constant) if other stabilizing influences (e.g., control system, nonlinear damping) become effective. If no such stabilizing effects occur, then a spin about the transverse axis (major axis) of the satellite will be approached resulting in the failure of the attitude control system [3].

3.9.1 Thrusting Maneuvers

Among the advantages of the dual-spin stabilized spacecraft are the velocity control capability and performance available by placing appropriate thrusters on the spinning portion. The effects of operating thrusters mounted on the rotor are much the same as in single body spin stabilized spacecraft. However, the presence of the despun platform can result in the combined center of mass being significantly offset from the bearing axis (Figure 3.8). Under these conditions, the operating of a radial jet (spin-synchronous thrust pulses transverse to the bearing axis) produces a torque about the vehicle's mass center, resulting in a change in the system angular momentum. If the platform despin attitude is constrained because of operational requirements, then the change in angular momentum is manifested as a change in rotor spin rate which simply must be accepted. However, if the platform attitude can be temporarily offset around the spin axis during the maneuver, then the magnitude and direction of the spin rate change can be controlled by rotating the platform prior to thruster firing to a fixed position of the mass center relative to the line of thrust. For example, if the platform center of mass can be placed in the plane of the average transverse thrust direction, nominally no change in relative spin occurs [5].

3.9.2 Conditions Leading to Nutation Growth

The action of a despin motor and/or bearing friction produces an interbody torque directed along the bearing axis of a dual-spin spacecraft. In general, this torque will change the spin rates of both bodies. It has been observed that as these spin rates change, it is possible to encounter situations where marked nutation growth occurs (θ increasing). There are several distinct causes or explanations for this growth. One involves consideration of internal energy dissipation. Internal dissipation has been recognized as a potential problem since the advent of the dual-spin concept. In addition to dissipation, nutation growth can result from (1) an unbalance on one body, (2) an unbalance on one body com-

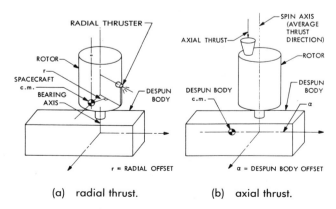

Figure 3.8 Thrusting maneuvers for nonsymmetrical dual-spin configurations [5].

bined with assymmetry on the other body, and (3) an assymmetry on one body.

Analyses of nutation and other related effects from static and dynamic unbalance are presented in references 6 through 10. Recovery from flat (transverse axis) spin employing wheel torque motor is discussed in references 11–13.

3.10 Recommended Practice

With respect to dual-spin stabilization, the following practices and considerations are recommended:

1. Energy dissipation of the spacecraft should be managed as follows:
    a. Highest spin rate members should be structurally stiff.
    b. Sloshing of fluids should not occur at sensitive frequencies (such as near the spin rate minus the nutational rate).
    c. A dissipative damper should be included on the more slowly spinning part to ensure stability of spin about the minor principal axis of the spacecraft.

2. The center of mass of the spinner should be as close to the bearing axis as possible.

3. The bearing axis should be the principal axis of the spinning part to prevent forced oscillations and nutation due to center of mass offset and cross products of inertia of the spinner.

4. The spinner should be dynamically symmetric (equal transverse moments of inertia) or the stability of spin about the minor principal axis of the spacecraft must be reevaluated by numerical simulation or a Floquet-type of analysis.

5. For the general case when the despun body has significant cross products of inertia or if its center of mass is not on the bearing axis, simulation of the system equations is recommended.

6. In addition to the above, the following are also recommended:
    a. Evaluate the effects of all energy dissipators on both the spinning and the despun parts.
    b. Evaluate the effects of bearing flexibility.
    c. Evaluate the effects of ''spinning up'' the despun section during apogee boost maneuvers.
    d. Maintain the center of mass on the thrust axis.

3.11 References

1. Iorillo, A. J. ''Hughes Gyrostat System,'' Proceedings of the Symposium on Attitude Stabilization and Control of Dual-Spin Spacecraft. The Aerospace Corporation TR-0159(3307–01)-16, November 1967.
2. Likins, P. W. ''Attitude Stability of Dual-Spin Spacecraft,'' Journal of Spacecraft and Rockets, Vol. 4, 1967, pp. 1638–1643.
3. Mingori, D. L. ''Effects of Energy Dissipation on the Attitude Stability of Dual-Spin Satellites,'' AIAA Journal, Vol. 7, 1969. pp. 20–27.
4. Vigneron, F. R. ''Motion of a Freely Spinning Gyrostat Satellite with Energy Dissipation,'' Astronautics Acta, Vol. 16, 1971.
5. ''Spacecraft Attitude Control During Thrusting Maneuvers,'' NASA SP-8059, February 1971.
6. Rimrott, F. P. G. Introductory Attitude Dynamics, Springer-Verlag 1989.
7. Adams, G. J. ''Dual-Spin Spacecraft Dynamics During Platform Spinup,'' Journal of Guidance and Control, Vol. 3, No. 1, January to February, 1980, pp. 29–36.
8. Tsuchiya, K. ''Attitude Behavior of a Dual-Spin Spacecraft Composed of Asymmetric Bodies,'' presented at the 7th Communications Satellite Systems Conference, San Diego, CA, 23–27 April 1978.
9. Harrison, J. A. ''Dynamics Problems Encountered During Platform Despin of a Dual-Spin Spacecraft with Asymmetric Rotor,'' Hughes Aircraft Co. Interdepartmental Correspondence Ref. No. 4091.2/628, 9 December 1976.
10. Scher, M. P., and R. L. Farrenkopf. ''Dynamics Trap State of Dual-Spin Spacecraft,'' AIAA Journal, Vol. 12, No. 12, December 1974, pp. 1721–1725.
11. Guelman, M. ''On Gyrostat Dynamics and Recovery,'' Journal of Astronautical Sciences, Vol. 37, No. 2, April-June 1989, pp. 109–119.
12. Guelman, M. ''Gyrostat Trajectories and Core Energy,'' Journal of Guidance, Control and Dynamics, Vol. 11, No. 6, Nov.-Dec. 1988, pp. 577–583.
13. Hubert, C. ''Spacecraft Attitude Acquisition from an Arbitrary Spinning or Tumbling State,'' Journal of Guidance and Control, Vol. 4, No. 2, Mar.-Apr. 1981, pp. 164–171.

Chapter 4
Three Axis Active Control

4.1 *Pure Jet Systems*

Three axis active control systems consist of two main classes: (1) mass expulsion (pure jet) and (2) angular momentum exchange (reaction wheel and control moment gyro) systems.

The basic advantages of a three axis active control system are: fast response, good pointing accuracy, possibility of using Sun oriented arrays (solar cells), and noninertial pointing capability. The disadvantages are: limited life due to fuel expenditure, scanning sensors, and no graceful performance degradation in the event of gas jet failure.

The addition of momentum exchange devices prolongs the life of the system at the expense of increased complexity.

For economy of operation, the jet thrust should be as small as possible. This is not always possible, however, especially if large initial angular velocities are imparted to the satellite. The larger the inertia of the satellite, the larger the required jet size as can be seen by equating initial angular momentum to impulse available.

Similarly, the limit cycle gas consumption is directly proportional to the vehicle inertia for the case of zero or very small external disturbance. In this case the time average propellant consumption rate is obtained by finding the ratio of propellant expended during one limit cycle to the limit cycle period.

Control torques in active attitude control systems are generally obtained from cold or hot gas and/or electric propulsion. The simplest cold gas systems use an inert gas stored in a high-pressure vessel with initial pressures up to 400 atmospheres. Normally the gas is passed through one or more pressure regulators so that the thrusters operate at nearly constant pressure. Thrust range is typically between 0.05 to 22 N (see Figure 4.1). Efficient operation can be achieved with pulse durations of less than 10 milliseconds to several seconds. The specific impulse I_{sp} can vary from 60 to 290 or more seconds, depending on the type of gas used. Two different types of propulsion systems are shown in Figures 4.2 and 4.3, with an example of a low-level thruster assembly shown in Figure 4.4. An example of a three axis (or body) stabilized spacecraft is shown in Figure 4.5.

4.2 *Typical Sequence of Control Operations*

After separation from the booster in the mission orbit (or an intermediate transfer orbit), the spacecraft must remove all tip-off rates (typical rates are 1 degree per second about all axes) and establish either a slow or a fast rotation, depending on whether further orbital maneuvers are required. A typical sequence of control operations is:

1. Acquire Sun.
2. Acquire Earth and star.
3. Establish normal limit cycle.
4. Perform reorientation.
5. Perform stationkeeping.

4.3 *Acquisition Approaches*

The Sun acquisition, or orienting the vehicle normal to the Sun line, can be achieved with the use of larger (5 to 22 N) thrusters and a set of Sun sensors. Also required may be a three axis rate gyro assembly. To ensure the capability of acquiring the Sun from any attitude, the Sun sensing must be over a full sphere (4π steradian coverage). A minimum of four sensors, each with a ± 70 degree field of view, may be required.

The angular acceleration required to initiate or stop rotation about a reference line can be determined by considering the angular impulse equation

$$I\dot{\theta} = F\ell\Delta t$$
$$\dot{\theta} = \frac{F\ell\Delta t}{I}$$
$$= \alpha\Delta t \qquad (4.1)$$

where α is the angular acceleration and I and $\dot{\theta}$ are the moment of inertia and the desired angular velocity. $F\ell$ and Δt are the applied torque and thrusting duration time, respectively. The angle θ traversed during torque application is given as

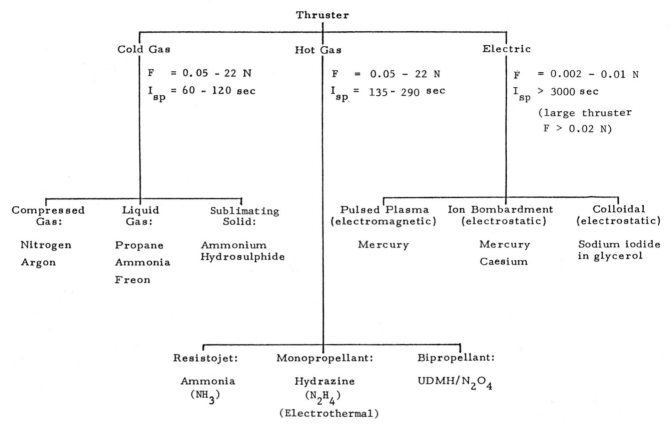

Figure 4.1 Mass expulsion thrusters.

$$\theta = \frac{1}{2} \alpha (\Delta t)^2$$
$$= \frac{1}{2} \alpha \left(\frac{\dot{\theta}}{\alpha} \right)^2 \qquad (4.2)$$

which can be solved to yield

$$\dot{\theta} = \sqrt{2\alpha\theta} . \qquad (4.3)$$

Earth acquisition can be achieved by initiating and stopping a slow rotation about the Sun line which permits the horizon scanners to intersect the Earth twice per orbit. Star acquisition may then be initiated, if needed. A typical functional attitude control system block diagram is shown in Figure 4.6.

4.4 On-Orbit Operation

The on-orbit operation can be established following the Earth/Sun/star acquisition. The attitude control may be with respect to either an inertially fixed direction or an Earth pointing vertical. The attitude angles about each of the three orthogonal axes (usually denoted as pitch, roll, and yaw) are required to stay within prescribed limits (deadband). The angular velocity within the deadband may also be specified but is required to be as small as possible to

prevent interference with the vehicle payload functions (observations, photography, etc.) and to minimize fuel consumption.

For the case of no external torques acting on the satellite, the single axis phase plane diagram is shown in Figure 4.7. The idealized limit-cycle operation of Figure 4.7 indicates that a constant angular rate $\dot{\theta}_o$ exists between the negative $(-\theta_d)$ and positive (θ_d) values of the deadband. Minimization of this velocity results in the improved gas consumption.

The time averaged propellant consumption rate can be obtained by finding the ratio of propellant expended during one limit cycle to the limit cycle period. Thus, for example, the total angular momentum change is $\Delta h = 2I(\Delta\dot{\theta})$, where $\Delta\dot{\theta} = 2\dot{\theta}_o$ and the propellant mass expended is $\Delta h / g I_{sp} \ell$, where

$I =$ satellite moment of inertia (kg-m^2)
$\dot{\theta}_o =$ limit cycle coast rate (rad/s)
$\theta_d =$ limit cycle dead zone (rad)
$I_{sp} =$ propellant specific impulse (s)
$\ell =$ jet moment arm about the vehicle center of mass (m)
$g =$ gravitational acceleration (m/s^2) .

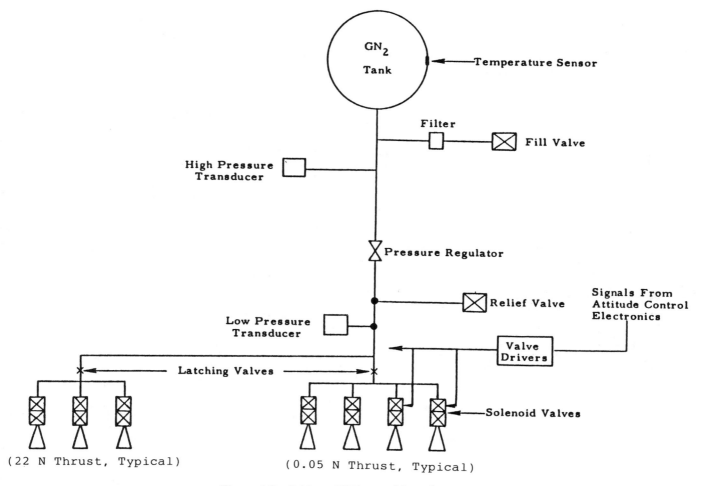

Figure 4.2 Cold gas (GN_2) propulsion subsystem.

The ideal propellant mass consumption rate in kg/s is then $\dot{m}$ where

$$\dot{m} = \frac{1}{P} \cdot \frac{\text{propellant mass}}{\text{cycle}}$$

$$= \frac{1}{P} \cdot \frac{2I(\Delta\dot{\theta})}{gI_{sp}\ell} \qquad (4.4)$$

where the period of the limit cycle is

$$P = \frac{4\theta_d}{\dot{\theta}_o} = \frac{8\theta_d}{\Delta\dot{\theta}}. \qquad (4.5)$$

In terms of jet thrust F and its duration Δt, the momentum change (torque impulse) per pulse is $F\ell\Delta t$. The ideal propellant consumption rate for the bang-bang-type operation may then be expressed in the form

$$\dot{m} = \frac{(F\ell\Delta t)^2}{4IgI_{sp}\ell\theta_d}. \qquad (4.6)$$

It can be seen from equation 4.6 that the propellant consumption rate is minimized for a small torque impulse and a large deadband.

In the presence of disturbances, the economic use of jet impulses implies that the jets should thrust only in opposition to the instantaneous disturbing torque whenever the magnitude of the torque exceeds a small fraction (0.1, say) of its maximum value. Mechanical efficiency considerations impose lower limits on the thrust F and pulse duration Δt which can be obtained from a jet. The control acceleration is $\alpha = F\ell/I$, and the nominal angular velocity increment from a jet is $\Delta\dot{\theta} = \alpha\Delta t$. If T_{min} is the minimum value of disturbing torque required when the jets thrust only in opposition to the disturbing torque, then the highest precision of control which can be achieved is illustrated in Figure 4.8.

The disturbance acceleration corresponding to T_{min} is $\alpha_{min} = T_{min}/I$. The required deadband is defined by $\theta_d = \alpha^2\Delta t^2/16\,\alpha_{min}$, and the error rate is $\dot{\theta} = \alpha\Delta t/2$. Under these conditions, a system of optimum design would obey Figure 4.8.

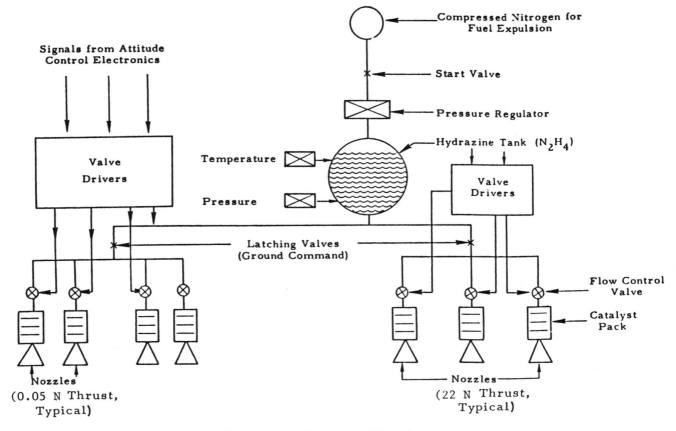

Figure 4.3 Hydrazine propulsion subsystem.

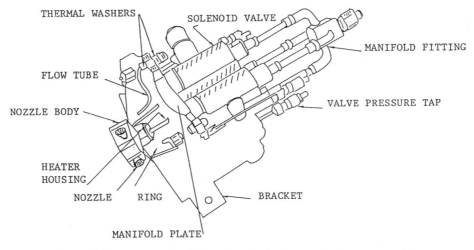

Figure 4.4 An example of a low-level attitude control thruster assembly.

4.5 Servomechanisms

Servomechanisms, or servos, are closed loop control systems used to determine the position, velocity, or acceleration of mechanical loads. The term *open loop* refers to systems for which the output variable is not measured nor is this information fed back to affect the system operation. In a closed loop system, however, the output variable is measured and compared with the input command; the difference between the two (the error) is used to make the

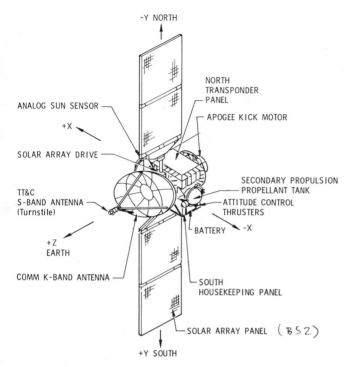

Figure 4.5 An example of a three axis stabilized spacecraft (B52) [9].

system correct itself. The material presented in this chapter is based largely on the developments found in references 1 through 7.

The basic properties of servos are revealed through the use of linear differential equations with constant coeffi-

cients. Laplace transforms are convenient tools which provide information and give some valuable areas of system insight. Laplace transforms convert linear differential equations having constant coefficients into simple algebraic equations.

4.5.1 Laplace Transforms

After the differential equations of the control system have been formulated and the initial conditions defined, the solution, by means of the Laplace transforms, consists of the following steps:

1. Transforming the differential equations with given initial conditions.
2. Solving for the transform of the desired unknown.
3. Performing the inverse transformation to obtain the solution in the time domain.

In the analysis of control systems, the differential equations are written in terms of the system parameters and the system variables and their time derivatives. The transformation of the equations is performed by introducing a change in the independent variable from time (t) to a complex variable (s) so chosen that derivatives, integrals, and trigonometric functions can all be expressed algebraically. This procedure is illustrated in the sections that follow.

4.5.2 Basic Properties

An important property of Laplace transforms is that a time dependent function has corresponding to it (and associated with it) a Laplace transform. This Laplace transform is essentially a function in the frequency domain. The Laplace

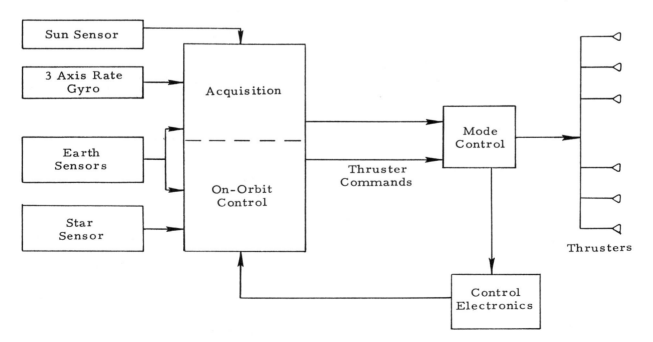

Figure 4.6 A typical functional attitude control system block diagram.

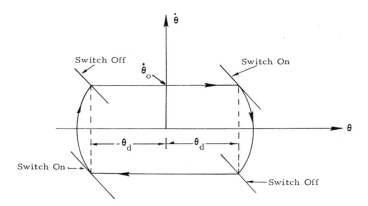

Figure 4.7   Single axis phase plane diagram.

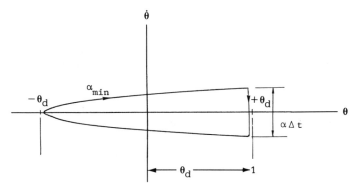

Figure 4.8   Ideal limit cycle with a minimum disturbing torque has a deadband $\theta_d = \alpha^2 \Delta t^2 / 16\alpha_{\min}$.

transform results from an operation on a function of time $f(t)$; this yields $F(s)$, which is a function of the complex variable $s = \sigma + j\omega$, where σ and ω are the real and imaginary parts of s, respectively.

Equation 4.7 gives the definition of the Laplace transform function $F(s)$ of the original time function $f(t)$; namely,

$$F(s) = \mathcal{L}[f(t)] = \int_0^\infty f(t)e^{-st}\, dt \ . \qquad (4.7)$$

4.5.3   Transform Examples

Consider the Laplace transform of a unit step function defined as follows:

$$f(t) = 1 \qquad \text{for } t > 0$$
$$\quad = 0 \qquad \text{for } t \le 0 \ . \qquad (4.8)$$

The unit step function has a value of one over the time range of $t > 0$ to $t = \infty$; therefore, its integral becomes the time integral of e^{-st} from zero to ∞. The Laplace transform for this case is particularly simple; namely

$$\mathcal{L}[f(t)] = \int_0^\infty 1 e^{-st}\, dt = \left| \frac{1}{-s} e^{-st} \right|_0^\infty$$
$$= \frac{1}{s} \ . \qquad (4.9)$$

This result shows that any unit step function applied to a circuit or to a servo may be expressed as the Laplace function K/s, where K is the magnitude of the step.

The Laplace transform of the simple decaying exponential $f(t) = e^{-at}$ is determined in equation 4.10. The direct evaluation of this integral yields

$$\mathcal{L}[f(t)] = \mathcal{L}(e^{-at})$$
$$= \int_0^\infty e^{-at}e^{-st}\, dt$$
$$= \int_0^\infty e^{-(a+s)t}\, dt$$
$$= \left. \frac{e^{-(a+s)t}}{-(a+s)} \right|_0^\infty$$
$$= -\frac{1}{-(a+s)} = \frac{1}{s+a} \ . \qquad (4.10)$$

It is also important that system stability may be readily determined from the form of the Laplace transform solution. Consider the time function e^{-at}. The transform of e^{-at} has been shown to be $1/(s + a)$; clearly "a" must be a positive quantity to result in a decaying exponential; a system characterized by a decaying exponential is stable.

The Laplace transforms of many functions have been tabulated. For simple functions, the Laplace transform integral can be evaluated using integrations by parts. A short table of transforms is shown in Table 4.1.

4.6   The Standard Diagram

A standard diagram for a single loop servo with unity feedback is illustrated in Figure 4.9, where K is a constant and $G(s)$ is a function of the complex variable s.

Table 4.1   List of Laplace Transforms

$f(t)$	$\mathcal{L}[f(t)] = \int_0^\infty e^{-st} f(t)\, dt$
1	$1/s$
e^{at}	$1/s - a$
$t^n, \ (n = 1, 2, \ldots)$	$n!/s^{n+1}$
$t^n e^{at}, \ (n = 1, 2, \ldots)$	$n!/(s-a)^{n+1}$
$\sin \omega t$	$\omega/(s^2 + \omega^2)$
$\cos \omega t$	$s/(s^2 + \omega^2)$
$\sinh \omega t$	$\omega/(s^2 - \omega^2)$
$\cosh \omega t$	$s/(s^2 - \omega^2)$
$e^{-at} \sin \omega t$	$\omega/[(s + a)^2 + \omega^2]$
$e^{-at} \cos \omega t$	$s + a/[(s + a)^2 + \omega^2]$
$\sqrt{t}$	$(\sqrt{\pi}/2)s^{-3/2}$
$1/\sqrt{t}$	$\sqrt{\pi/s}$
Unit-impulse function or delta function $\delta(t - t_o)$	$\mathcal{L}[\delta(t - t_o)] = e^{-st_o}$ for $t_o \ge 0$ $\mathcal{L}[\delta(t)] = 1$

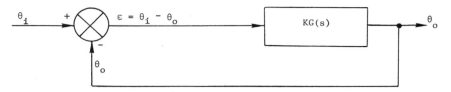

Figure 4.9   Standard diagram for a servo with unity feedback.

4.6.1   Single Loop Servo with Unity Feedback

The input to the system is θ_i, which is the reference variable. The output is θ_o, the controlled variable, and ε is the error. The error multiplier times the loop gain (direct transfer function) $KG(s)$ gives the controlled variable output θ_o. This is, in turn, "fed back" negatively and compared with the input variable θ_i to form the error ε.

The following definitions are given:

Closed loop transfer function

$$= \frac{\theta_o}{\theta_i} = \frac{KG(s)}{1 + KG(s)} \quad (4.11)$$

Open loop transfer function $= \dfrac{\theta_o}{\theta_i - \theta_o} = KG(s)$

$$= \frac{\theta_o}{\varepsilon} \quad (4.12)$$

where $\theta_i - \theta_o$ is the difference or error function ε.

The closed loop transfer function represents the output signal divided by the input signal. It is obtained from the definition of the error function ε, that is

$$\varepsilon = \theta_i - \theta_o \quad (4.13)$$

and therefore

$$\theta_o = KG(s)(\theta_i - \theta_o) \quad (4.14)$$

from which

$$\frac{\theta_o}{\theta_i} = \frac{KG(s)}{1 + KG(s)}$$
$$= \text{closed loop transfer function} . \quad (4.15)$$

Similarly,

$$\varepsilon = \frac{\theta_o}{KG(s)}$$

$$\theta_o = \frac{KG(s)\theta_i}{1 + KG(s)} \quad (4.16)$$

and therefore

$$\frac{\varepsilon}{\theta_i} = \frac{1}{1 + KG(s)}$$
$$= \text{error transfer function} . \quad (4.17)$$

The limiting values of the closed loop and the error transfer functions for various values of the open loop gain (transfer function) are illustrated in Table 4.2 [1].

The infinite gain in Table 4.2 results when $KG(s) = -1$ or the open loop phase (defined in section 4.11.1) is 180 degrees.

It can be seen from Table 4.2 that good error performance (small errors) results when the open loop gain is large. However, as the open loop gain increases, instability may result. Servo stability therefore occurs only over a specified range. This can be determined by analytical or graphical methods of analysis.

4.6.2   Servo with Nonunity Feedback

Figure 4.10 illustrates a servo with nonunity feedback or the inclusion of a transfer function in the return feedback line.

The following definitions are given:

$$\text{Direct transfer function} = KG(s)$$
$$\text{Open loop transfer function} = KG(s)H(s)$$
$$\text{Overall (closed loop) transfer function}$$
$$= \frac{\theta_o}{\theta_i} = \frac{KG(s)}{1 + H(s)KG(s)} .$$

The feedback transfer function $H(s)$ may be selected to achieve a desired transient and/or steady-state performance of the servo. It compensates for the undesirable characteristics of the direct (open loop) transfer function and is a valuable servo design tool.

Now consider more than one block (transfer function) in a signal path as is illustrated in Figure 4.11.

Table 4.2   Limiting Values of Closed Loop and Error Transfer Functions

Open Loop Gain	Closed Loop Gain	Error Gain
$KG(s)$	$\dfrac{KG(s)}{1 + KG(s)}$	$\dfrac{\varepsilon}{\theta} = \dfrac{1}{1 + KG(s)}$
$\gg 1$	1	$1/KG(s)$
$\ll 1$	$KG(s)$	1
1	$1/2$ to ∞	$1/2$ to ∞

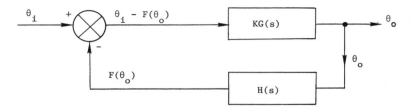

Figure 4.10 Single loop servo with nonunity feedback.

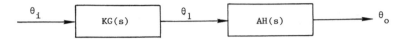

Figure 4.11 Two series blocks or transfer functions in a signal path.

The first block has an input θ_i and an output θ_1; therefore

$$\frac{\theta_1(s)}{\theta_i(s)} = KG(s) .$$

Similarly, θ_1 serves as input to the second block with output θ_o and

$$\frac{\theta_o(s)}{\theta_1(s)} = AH(s) .$$

Hence, the transfer function of the series combination is

$$KG(s)AH(s) = \frac{\theta_o(s)}{\theta_i(s)} . \tag{4.18}$$

4.7 A Simple Servomechanism

Now consider the same simple servo with open loop gain of ω_c and apply to it $F_{in}(t)$, a step function of unit magnitude as illustrated in Figure 4.12.

Equation 4.19 expresses the output controlled variable $\theta_o(s)$ in terms of the input function and the closed loop transfer function of the servo. That is, the output transform is the product of the transform of the input function and the transfer function,

$$\theta_o(s) = F_{in}(s)G'(s) . \tag{4.19}$$

The transform of the input function $F_{in}(t)$ is given in equation 4.20. By taking the product of the right side of equations 4.20 and 4.21, the transform of the output signal $\theta_o(s)$ is determined in equation 4.22; that is,

$$F_{in}(s) = \frac{1}{s} \tag{4.20}$$

and

$$G'(s) = \frac{\omega_c}{s + \omega_c} . \tag{4.21}$$

Therefore,

$$\theta_o(s) = \frac{1}{s} \times \frac{\omega_c}{s + \omega_c} \tag{4.22}$$

The system time response (transient response) can now be determined by taking the inverse transform of $\theta_o(s)$. The controlled variable in Laplace transform form is given in equation 4.22. This result is restated in equation 4.23 in "partial fraction" form. It is necessary here to break the transform into two parts to obtain the transient solution

$$\frac{1}{s} \times \frac{\omega_c}{s + \omega_c} = \frac{A}{s} + \frac{B}{s + \omega_c} . \tag{4.23}$$

A partial fraction solution may be obtained by finding values of the constants A and B that satisfy equation 4.23 for

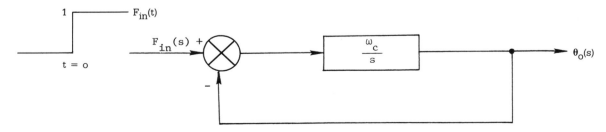

Figure 4.12 Single integrator in a loop (standard form).

all values of s. To do this, the left and right sides of equation 4.23 are next multiplied by s; this gives

$$\frac{\omega_c}{s + \omega_c} = A + \frac{Bs}{s + \omega_c} . \qquad (4.24)$$

By letting s go to zero, the value of A is immediately found to be 1. This is a valid procedure because equation 4.23 must hold for all values of s. Similarly, equation 4.23 may be multiplied by $s + \omega_c$ to provide equation 4.25

$$\frac{\omega_c}{s} = \frac{A(s + \omega_c)}{s} + B . \qquad (4.25)$$

If $s = -\omega_c$, then equation 4.25 yields $B = -1$. Substituting the values $A = 1$ and $B = -1$ into equation 4.25 yields

$$\frac{1}{s} \times \frac{\omega_c}{s + \omega_c} = \frac{1}{s} - \frac{1}{s + \omega_c} = \theta_o(s) . \qquad (4.26)$$

Taking the inverse transform of $\theta_o(s)$ gives

$$\theta_o(t) = 1 - e^{-\omega_c t} . \qquad (4.27)$$

The inverse transform therefore gives the output as a function of time which is shown in Figure 4.13.

The above solution was relatively straightforward and simple; the important results are (1) that a servo acts as a low pass filter and (2) that the output rise time due to a step function input is $1/\omega_c$ where ω_c is the gain of the integrator. In one period of $1/\omega_c$ the output rises to a value of $1 - 1/e$ or approximately 63 percent of the final value.

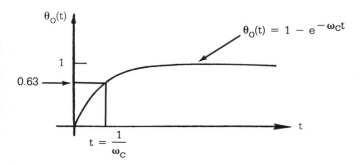

Figure 4.13 Step function response of a single integrator in the loop.

4.8 Stability

The stability of servomechanisms can be studied using frequency response plots, and the system transient response can be obtained from the roots of the characteristic equation. These techniques are developed in the following section.

4.8.1 Poles and Zeros

Figure 4.14 shows the basic servo block diagram for a system employing complete feedback. Equation 4.28 gives an assumed expression for $G(s)$. The Laplace variables is a complex quantity consisting of a real part sigma (σ) and an imaginary part ($j\omega$) as given in equation 4.29. That is,

$$G(s) = \frac{k(s + b)}{s(s + a)} \qquad (4.28)$$

$$s = \sigma + j\omega . \qquad (4.29)$$

Two dimensions are required to specify s [real (Re) and imaginary (Im)], and a third dimension, therefore, is needed to plot the magnitude of $G(s)$. Examination of Figure 4.15 suggests how a three dimensional plot may be made for an expression such as equation 4.28. Equation 4.28 indicates that when $s = 0$, the denominator goes to zero and, therefore, the gain goes to infinity; this is shown in Figure 4.15 as a "pole" at the origin. It is also called a pole because the gain goes up vertically above the s plane very much as a pole rises above the ground. In this plot there are two poles, one at the origin and the other at $s = -a$. The latter pole is present because the gain expression also goes to infinity as $s = -a$.

As s is allowed to take on values along the imaginary axis, there is a different effect. For s equal to $j\omega$ (along the imaginary axis), the expression becomes $G(j\omega)$, and the frequency response is determined. It can be seen that when the frequency is zero ($j\omega = 0$), the gain function goes to infinity. Another example is shown in Figure 4.16; in this figure there are conjugate poles at $-c + j\omega_1$, and at $-c - j\omega_1$.

$$G(s) = \frac{s + b}{s(s + a)(s + c + j\omega_1)(s + c - j\omega_1)}$$

$$= \frac{s + b}{s(s + a)[(s + c)^2 + (\omega_1)^2]} . \qquad (4.30)$$

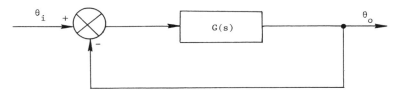

Figure 4.14 The basic or standard servo block diagram.

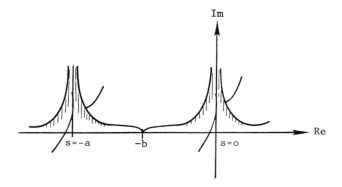

Figure 4.15 The three-dimensional plot of a complex function $k(s + b)/s(s + a)$.

In this case, as the frequency variable moves along the positive imaginary axis and approaches $s = +j\omega_1$, the gain becomes large in the region of resonance.

In summary, when s equals a denominator root, the gain goes to infinity and the system has a pole. When s equals a numerator root, the gain magnitude goes to zero and the system has a zero. When s moves along the imaginary axis near a complex pole, the gain in the region of resonance is determined.

Figure 4.17 is a simplified representation of Figure 4.16. This representation, instead of being a three-dimensional drawing, is reduced to two dimensions by indicating pole locations on the s-plane as crosses and zero locations as small circles. The magnitude plot can be pictured as coming up out of the paper; crosses indicate high magnitudes, and

circles indicate the magnitude dropping to zero at the surface of the paper.

There is a direct correspondence between pole position and transient performance. For instance, a pole at $-a$ corresponds to an e^{-at} response. If a pole is at $+a$ (to the right of the imaginary axis), this will give an e^{+at} term; this is an unstable response, since it implies a continually increasing output. Figure 4.18 shows the complex s-plane; stable poles are to the left of the imaginary axis and unstable poles are to the right of the imaginary axis.

4.9 S-Plane Quadratic Transfer Function Response

Consider Figure 4.19; this represents a single integration (a pole at the origin) and a pole in the left half-plane at $-\alpha$.

Equation 4.31 defines the output function in terms of the input and the closed loop gain. Namely,

$$\theta_o(s) = \theta_i(s)G'(s) \qquad (4.31)$$

where

$$
\begin{aligned}
G'(s) &= \frac{G(s)}{1 + G(s)} \\
&= \frac{k/s(s + \alpha)}{1 + [k/s(s + \alpha)]} \qquad (4.32) \\
&= \frac{k}{s^2 + \alpha s + k} .
\end{aligned}
$$

If an input step function is assumed, the output function is given by equation 4.33. This output function has three roots;

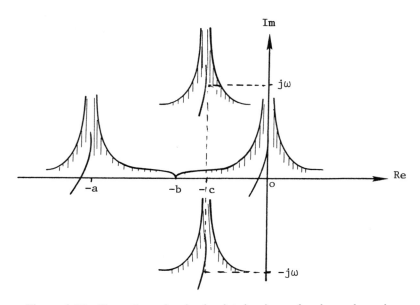

Figure 4.16 Three-dimensional gain plot showing real and complex poles.

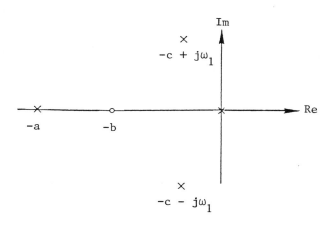

Figure 4.17 Two-dimensional representation of three-dimensional plot.

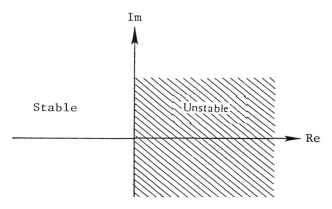

Figure 4.18 Complex s-plane showing regions for stable and unstable poles (denominator roots).

one root is at zero because of the input step function. Therefore,

$$\theta_o(s) = \frac{k}{s(s^2 + \alpha s + k)} . \quad (4.33)$$

The closed loop system has two conjugate poles or roots at $(-\alpha + \sqrt{\alpha^2 - 4k})/2$ and $(-\alpha - \sqrt{\alpha^2 - 4k})/2$.

The closed loop gain of equation 4.32 is in the form of a quadratic and lends itself to a generalized interpretation.

Equation 4.34 considers what is commonly called Case 1; this is for α^2 greater than $4k$.

$$\text{Case 1: } \alpha^2 > 4k . \quad (4.34)$$

For this case, the term under the square root sign is a positive number yielding, therefore, two distinct (different) real roots at $-p_1$, and $-p_2$. Equation 4.35 defines this output function in terms of the known poles; that is,

$$\theta_o(s) = \frac{k_1}{s} + \frac{k_2}{s + p_1} + \frac{k_3}{s + p_2} . \quad (4.35)$$

The k_1 term is for the input step function, the k_2 term is for the pole at $-p_1$, and the k_3 term is for the pole at $-p_2$. The resulting time function, given in equation 4.36, shows an exponential decay of the closed loop response:

$$\theta_o(t) = k_1 + k_2 e^{-p_1 t} + k_3 e^{-p_2 t} . \quad (4.36)$$

Figure 4.20 shows the two system poles on the s-plane; the pole at the origin is not shown because it is due to the applied step function input and is not due to the system itself.

Figure 4.21 and equations 4.37 and 4.39 consider Case 2.

$$\text{Case 2: } \alpha^2 = 4k \quad (4.37)$$

$$\theta_o(s) = \frac{k_4}{s} + \frac{k_5}{(s + \alpha/2)^2} + \frac{k_6}{s + \alpha/2} \quad (4.38)$$

$$\theta_o(t) = k_4 + k_5 t e^{-(\alpha/2)t,} + k_6 e^{-(\alpha/2)t} . \quad (4.39)$$

For this case, $\alpha^2 = 4k$ and the argument under the radical becomes zero; therefore, there are two identical real roots, both at $-\alpha/2$. These poles are shown in Figure 4.21; the resulting output function is given by equation 4.39. It will be noted that because there is a repeated root (double pole), one transient term is multiplied by t.

$$\text{Case 3: } \alpha^2 < 4k \quad (4.40)$$

$$\theta_o(s) = \frac{k_7}{s} + \frac{k_8}{s + \alpha - j\omega} + \frac{k_9}{s + \alpha + j\omega} \quad (4.41)$$

$$\theta_o(t) = k_7 + k_{10} e^{-\alpha t} \sin(\omega t + \phi) \quad (4.42)$$

For Case 3, α^2 is less than $4k$ and the quantity under the square root sign then becomes negative. As a result, the poles have imaginary as well as real parts, and by definition are complex. The root locations are at $-\alpha \pm j\omega$ and are

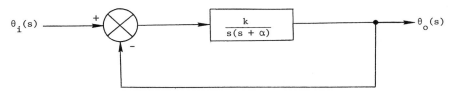

Figure 4.19 Standard diagram of a single integration and a lag term in the loop.

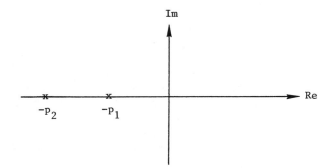

Figure 4.20   Case 1 ($\alpha^2 > 4k$): Two real roots.

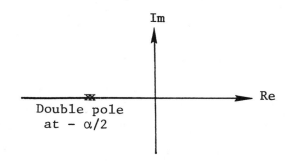

Figure 4.21   Case 2 ($\alpha^2 = 4k$): Two repeated real roots.

shown in Figure 4.22. The exponentially decaying sinusoidal output function is given by equation 4.42.

These three situations are the basic ones for a closed loop quadratic system.

4.10 Attitude Control System Example

The schematic diagram of the vehicle to be controlled is shown in Figure 4.23. Primary disturbances result from the firing of either of the two orbit injection solid propellant

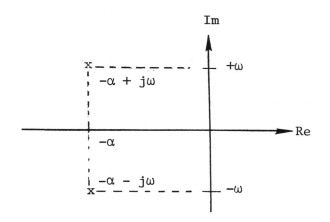

Figure 4.22   Case 3 ($\alpha^2 < 4k$): Two complex conjugate roots.

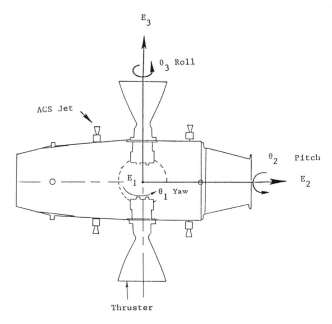

Figure 4.23   Vehicle configuration illustrating thruster control of pitch, roll, and yaw motions.

rockets. Pitch, roll, and yaw position and rate must be maintained close to those existing prior to the initiation of the injection burn.

The position (angle) and angular rates can be measured by rate and rate integrating gyros. The relationship between the gyro output signals, which are proportional to the angular rate ω, and the yaw (θ_1), pitch (θ_2), and roll (θ_3) body angles and their rates for a yaw, pitch, and roll sequence of rotations can be expressed as

$$\begin{pmatrix} \dot{\theta}_1 \\ \dot{\theta}_2 \\ \dot{\theta}_3 \end{pmatrix} = \begin{pmatrix} 1 & s\theta_1 c\theta_2 & c\theta_1 c\theta_2 \\ 0 & c\theta_1 & -s\theta_1 \\ 0 & s\theta_1/c\theta_2 & c\theta_1/c\theta_2 \end{pmatrix} \begin{pmatrix} \omega_1 \\ \omega_2 \\ \omega_3 \end{pmatrix} \quad (4.43)$$

where $s\theta_1 = \sin\theta_1$, $c\theta_1 = \cos\theta_1$, etc., and ω_1, ω_2, and ω_3 are the body components of the angular velocity (measured by a rate gyro). The $\dot{\theta}_1$, $\dot{\theta}_2$, and $\dot{\theta}_3$ components are the yaw, pitch, and roll rates of the vehicle. Integration of the latter results in the yaw, pitch, and roll angles (errors) which must be maintained within specified limits.

Combining the above kinematical relations with the uncoupled dynamical equation for angular motion (Euler's equation), a simplified single (e.g., yaw) channel reaction attitude control system can be obtained as is shown schematically in Figure 4.24. Both rate and position feedback signals are combined with a command signal θ_{CI} to produce a yaw control torque u_1. The latter is added to any disturbance torque T_{D1} which produces the yaw axis angular acceleration $\dot{\omega}_1$. Integrating $\dot{\omega}_1$ produces ω_1 which, combined with the components of ω_3 and ω_2, yields the desired $\dot{\theta}_1$ body rate. The integral of $\dot{\theta}_1$ yields the controlled variable θ_1, the yaw angle.

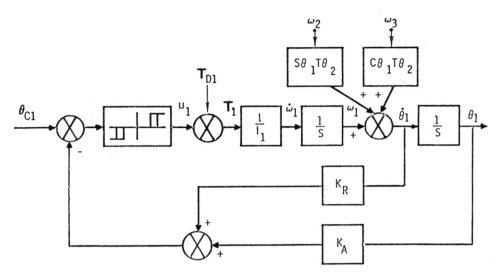

Figure 4.24 Simplified yaw channel reaction attitude control system.

4.10.1 Transfer Function Analysis

To determine the position K_A and rate K_R gains and study the performance of the system presented in Figure 4.24, a closed loop transfer function (TF) is required. Such a closed loop transfer function relating θ_1 to T_{D1} is shown in Figure 4.25 under the given simplifying assumptions and for the instantaneous application of the control torque u_1.

The poles of the following closed loop TF (with unit subscripts omitted for simplicity)

$$\frac{\theta(s)}{T_D(s)} = \frac{1}{Is^2 + K_R s + K_A} \quad (4.44)$$

are the roots s_1 and s_2 of the characteristic equation

$$Is^2 + K_R s + K_A = 0 . \quad (4.45)$$

That is,

$$s_{1,2} = \frac{-K_R \pm \sqrt{K_R^2 - 4IK_A}}{2I} .$$

The K_R, K_A values can be found if underdamped (oscillatory), critically damped, or overdamped response for the

dynamical system is considered. A simple but not optimal solution results if critical damping is assumed. Then

$$\sqrt{K_R^2 - 4IK_A} = 0 \quad (4.46)$$

and

$$s_{1,2} = -K_R/2I$$
$$= -\frac{1}{\tau} \quad (4.47)$$

where τ is the time constant. Also

$$K_A = K_R^2/4I . \quad (4.48)$$

Thus

$$\frac{\theta(s)}{T_D(s)} = \frac{\tau^2}{I(\tau s + 1)^2} . \quad (4.49)$$

4.10.2 Response to Step Disturbance Torque

The performance of the system can be analyzed by examining the time domain response to a step disturbance in equation 4.49. The Laplace transform of the output is

$$\theta(s) = \frac{\tau^2 T_D}{Is(\tau s + 1)^2} \quad (4.50)$$

where $T_D(s) = T_D/s$ and T_D is the magnitude of the step disturbance torque.

The time response of equation 4.50 is

$$\theta(t) = \frac{T_D}{I}[\tau^2(1 - e^{-t/\tau}) - \tau t e^{-t/\tau}] . \quad (4.51)$$

The maximum value (or steady-state value) of $\theta(t)$ occurs as time goes to infinity. In this case

$$\theta(t)_{max} \to \frac{T_D \tau^2}{I} . \quad (4.52)$$

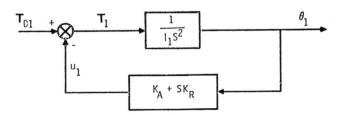

Figure 4.25 Approximation of yaw channel ACS.

The steady-state response of the system can also be obtained from the "final value theorem." This theorem relates the final value of $\theta(t)$, $\theta(\infty)$, to the Laplace transform $\theta(s)$ as follows:

$$\theta(\infty) = \lim_{s \to 0} s\theta(s) = \frac{T_D \tau^2}{I} .$$

It is valid, provided that $s\theta(s)$ is analytic on the imaginary axis and to the right of the imaginary axis; i.e., it applies only to stable systems.

Thus, if the maximum value for the controlled variable θ_{max} is specified, then the time constant τ can be obtained from equation 4.52 and is found to be

$$\tau = \sqrt{\frac{\theta_{max} I}{T_D}} . \qquad (4.53)$$

Also, solving equation 4.47 for K_R and using equation 4.53 to eliminate τ yields

$$K_R = \frac{2I}{\tau} = 2\sqrt{\frac{IT_D}{\theta_{max}}} . \qquad (4.54)$$

Similarly, from equation 4.48, we have

$$K_A = \frac{K_R^2}{4I} = \frac{T_D}{\theta_{max}} . \qquad (4.55)$$

The maximum control torque u which is required to limit θ to θ_{max} can be found from Figure 4.25 as

$$u(s) = -(K_A + sK_R)\theta(s) \qquad (4.56)$$

which, for a step disturbance M_D, becomes

$$u(s) = -\frac{(1 + 2\tau s)T_D}{s(\tau s + 1)^2} . \qquad (4.57)$$

Again, by the "final value" theorem

$$\lim_{t \to \infty} u(t) = \lim_{s \to 0} su(s) = -T_D \qquad (4.58)$$

or, in the limit, the control torque must be equal to the disturbance.

The maximum control torque u_{max}, however, can be found from the inverse Laplace transform of equation 4.57. Thus,

$$u(t) = -T_D \left[(1 - e^{-t/\tau}) + \frac{t}{\tau} e^{-t/\tau} \right] . \qquad (4.59)$$

The maximum of equation 4.59 occurs when $t = 2\tau$. Then

$$u(t)_{max} = -T_D(1 + e^{-2}) . \qquad (4.60)$$

4.11 Graphical Methods

Root locus and frequency response plots can be applied to analyze and design a control system. The frequency response plots of the system's open loop transfer function contain information about the behavior of closed loop systems. These plots can be easily generated, and they permit proper control gain to be selected. This method is most useful, because it does not require the values of the characteristic roots which may be difficult to obtain for higher order systems. The root locus and frequency response plots can be sketched by hand with the aid of a calculator, or they can be adapted to computer use.

The decibel is a generally accepted unit which describes the magnitude of a number. Thus, if n is the value in decibels of the number N, the two are related as

$$n = 20 \log_{10} N . \qquad (4.61)$$

A ratio between two numbers N_1 and N_2 in decibels is given as

$$\frac{n_1}{n_2} = 20 \log_{10} N_1 / N_2 . \qquad (4.62)$$

4.11.1 Nyquist Diagram

The practical aspects of the Nyquist stability theory (as described in References 4–7, for example) may best be illustrated by plotting the frequency response of the open loop transfer function on a polar graph known as a Nyquist Diagram.

The frequency response of a system represents the output of the system when the input is sinusoidal in form. The system can thus be described by the output-input amplitude ratio and phase difference over the entire range of forcing frequency from zero to infinity. The steady-state sinusoidal output-input amplitude ratio for an open loop transfer function can be obtained by substituting $j\omega$ for s in the system transfer function $G(s)$. The $G(j\omega)$ transfer function is now a complex number, the amplitude and phase of which may be plotted on a complex plane having real and imaginary axes. The resulting curve is called the polar or Nyquist plot.

Consider, for example, the open loop transfer function $G(s)$ whose frequency response $G(j\omega)$ can be expressed in the form of

$$G(j\omega) = A(\omega)e^{j\theta(\omega)} \qquad (4.63)$$

where the amplitude is

$$A(\omega) = \sqrt{Re^2 + Im^2} \qquad (4.64)$$

and the phase angle is

$$\theta = \tan^{-1}\left(\frac{Im}{Re}\right) . \qquad (4.65)$$

Here Re and Im represent the real and imaginary parts of $G(j\omega)$.

For example, consider the following transfer function for an integrator:

$$G(s) = k/s . \qquad (4.66)$$

Let $s = j\omega$, then

$$G(j\omega) = k/j\omega \qquad (4.67)$$

and

$$A(\omega) = |G(j\omega)|$$
$$= k/\omega$$
$$\theta = -90° \qquad (4.68)$$

Thus, the amplitude in decibels is

$$A(\omega) = 20 \log k - 20 \log \omega \qquad (4.69)$$

The attenuation with frequency of the integrator transfer function is therefore 6 dB per octave or 20 dB per decade.

As another example, consider the open loop transfer function

$$G(s) = k/(1 + Ts) . \qquad (4.70)$$

Let $s = j\omega$, then

$$G(j\omega) = k/(1 + j\omega T)$$
$$|G(j\omega)| = k/\sqrt{1 + (\omega T)^2} \qquad (4.71)$$

and

$$\theta = \tan^{-1}(-\omega T) . \qquad (4.72)$$

Thus, the amplitude in decibels is

$$A(\omega) = 20 \log k - 20 \log \sqrt{1 + (\omega t)^2} . \qquad (4.73)$$

A complex plane plot of $G(j\omega T)/k$ is illustrated in Figure 4.26.

Equation 4.63 can be plotted on a complex plane where it may, for example, appear as shown in Figure 4.27.

The closed loop system with unity feedback is then stable when the -1 point on the real axis in Figure 4.27 is to the left of the $G(j\omega)$ curve at a frequency of ω_p when the phase lag angle θ (measured clockwise from the real axis) is 180 degrees.

The gain margin, measured in decibels, for stability of the closed loop system is, in this case, equal to the amount of additional gain [amplitude of $G(j\omega)$] at the frequency of 180 degrees phase lag to cause the gain plot to pass through the -1 point when the closed loop gain goes to infinity. That is

$$\lim_{G(j\omega) \to -1} \left[\frac{G(j\omega)}{1 + G(j\omega)} \right] = \infty . \qquad (4.74)$$

For example, if $G(j\omega) = 0.8$, then the gain margin is GM $= 20 \log(1) - 20 \log(0.8) \approx 2$ dB.

The phase margin is the amount of additional phase lag required at the frequency of unity gain to cause the gain plot to pass through the -1 point and thus produce instability. Practical values of gain and phase are typically about 8 dB and 50 degrees.

4.11.2 Bode Diagram

The Bode diagram shown in Figure 4.28 plots open loop gain (in dB) versus log ω and phase angle versus log ω. The gain and phase crossover frequencies correspond to those shown in Figure 4.27 of the Nyquist diagram.

4.11.3 Nichols Diagram

The Nichols diagram plots the open-loop frequency response of a particular system versus phase lag angle θ on

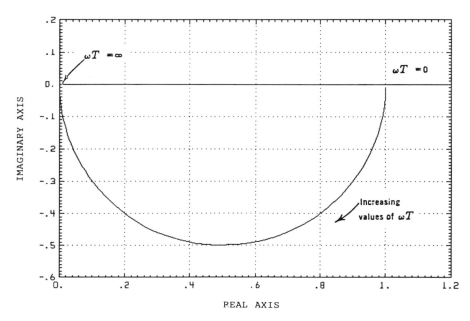

Figure 4.26   Nyquist plot of $G(j\omega T)/k$.

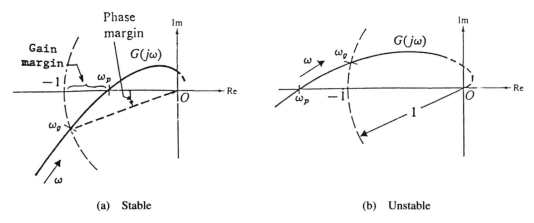

Figure 4.27   Nyquist diagram of $G(j\omega) = A(\omega)e^{j\theta(\omega)}$.

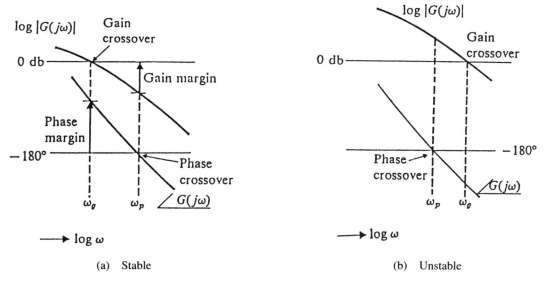

Figure 4.28   Bode diagram of $G(j\omega) = A(\omega)e^{j\theta(\omega)}$.

a chart with constant closed loop response (gain and phase). The gain and phase margins as well as the closed loop gain and phase are thus readily obtained from the Nichols chart, as is illustrated in Figure 4.29. The Nichols chart represents the graphical solution of $KG(s)/[1 + KG(s)]$ from $KG(s)$.

4.12   Time and Frequency Response of a Position Control Servo

Consider a motor controlled by a servo as illustrated schematically in Figure 4.30.

Let the block diagram denote an error channel where the amplifier current is proportional to the error ($\theta_i - \theta_o$). This generates a torque in the motor and load, producing a shaft angular acceleration of $d^2\theta_o/dt^2$ which overcomes the frictional (damping) losses which are assumed to be propor-

tional to the angular velocity. The governing equation is of the form

$$J \frac{d^2\theta_o}{dt^2} + B \frac{d\theta_o}{dt} = k(\theta_i - \theta_o) \qquad (4.75)$$

where J is the moment of inertia of the motor and load, B is the damping constant, and k is the torque proportionality factor. Equation 4.75 can be written as

$$\frac{d^2\theta_o}{dt^2} + \frac{B}{J} \frac{d\theta_o}{dt} + \frac{k}{J} \theta_o = \frac{k}{J} \theta_i \qquad (4.76)$$

or, in terms of more usual notation

$$\ddot{\theta}_o + 2\zeta\omega_n\dot{\theta}_o + \omega_n^2\theta_o = \omega_n^2\theta_i \qquad (4.77)$$

where $\omega_n = \sqrt{k/J}$ is the "natural frequency," and the damping ratio $\zeta = B/2\sqrt{kJ}$ is the ratio of the system damping to the critical damping.

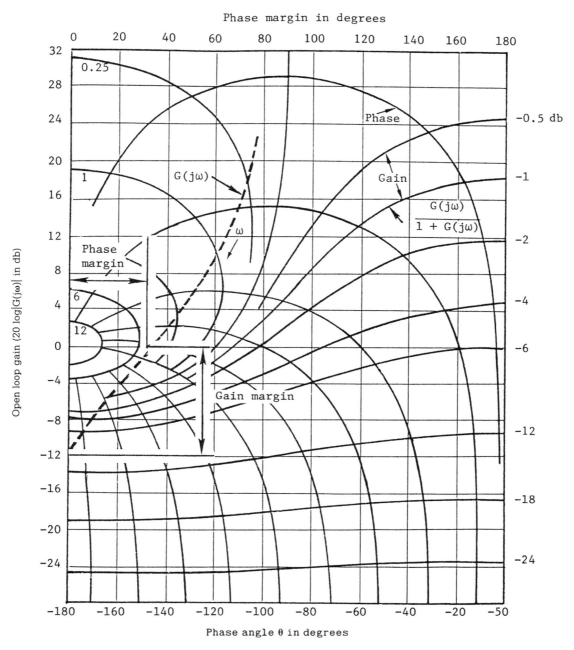

Phase margin in degrees

Open loop gain (20 log|G(jω)| in db)

Phase angle θ in degrees

Figure 4.29 Nichols chart for $G(j\omega)$.

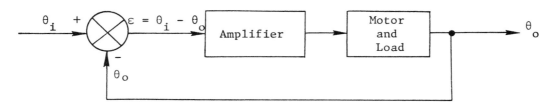

Figure 4.30 Schematic diagram of a servo-controlled motor.

In Laplace operational notation, equation 4.77 becomes

$$(s^2 + 2\zeta\omega_n s + \omega_n^2)\theta_o(s) = \omega_n^2\theta_i(s) \qquad (4.78)$$

where $\theta_o(s)$ and $\theta_i(s)$ are Laplace variables. The closed loop transfer function is of the form

$$\frac{\theta_o(s)}{\theta_i(s)} = \frac{\omega_n^2}{s^2 + 2\zeta\omega_n s + \omega_n^2}$$
$$= \frac{1}{(s/\omega_n)^2 + 2\zeta s/\omega_n + 1}. \qquad (4.79)$$

The roots (poles) of the closed loop transfer function are:

$$P_{1,2} = -\zeta\omega_n \pm \omega_n \sqrt{\zeta^2 - 1}$$
$$= -\zeta\omega_n \pm j\omega_n \sqrt{1 - \zeta^2}. \qquad (4.80)$$

For the general quadratic transfer function, the magnitude of the damping ratio determines the following three cases: (1) when $\zeta > 1$, the roots are real and distinct; (2) when $\zeta = 1$, the roots are real and identical; (3) when $\zeta < 1$, the roots are complex conjugates.

The response of the closed loop transfer function to a unit step input is illustrated in Figure 4.31, and the peak overshoot above unity is given in Figure 4.32 as a function of the damping ration ζ. The response to a sinusoidal input of frequency ω is illustrated in Figure 4.33, which also shows the corresponding phase angle measured negatively (clockwise) from the positive real axis in the complex plane.

4.13   The Root-Locus Method

The root-locus method is also largely graphical but makes use of the pole and zero concept instead of logarithmic plots.

The poles (infinities or denominator zeros) and zeros of the open loop transfer function are plotted on the $s = \sigma + j\omega$ plane. Since $KG(s)$ is the ratio of two polynomials

$$KG(s) = \frac{P_1(s)}{P_2(s)}$$
$$= \frac{(s - r_1)(s - r_2)\ldots(s - r_m)}{(s - a_1)(s - a_2)\ldots(s - a_n)} \qquad (4.81)$$

this involves factoring $KG(s)$ and plotting all r_i and a_i. Ultimately, it is necessary to find poles of the closed loop transfer function $\theta_o(s)/\theta_i(s) = KG(s)/[1 + KG(s)]$. Poles of this are the zeros of $1 + KG(s)$ and will occur where the phase angle of $KG(s)$ is 180 degrees and its magnitude is 1. Therefore, the root loci are plots of the variations of the poles of the closed loop system function with changes in the open loop gain.

The root loci thus constitute a graphical method for the approximate determination of the zeros of a polynomial. The usefulness of the root loci in the design of feedback control systems depends directly on the ease with which the loci can be constructed. In a design problem, the general shapes of the loci are sketched approximately for a number of different sets of open loop poles and zeros; with a few

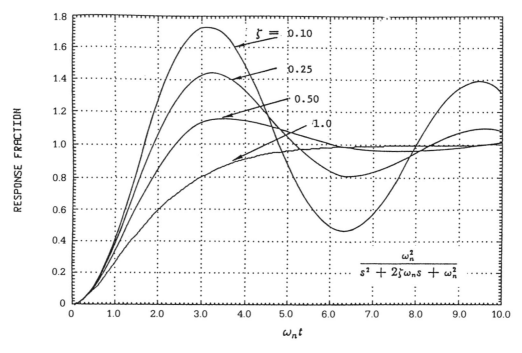

Figure 4.31   Unit-step-function responses for a damped oscillator for various damping ratios, ζ [8].

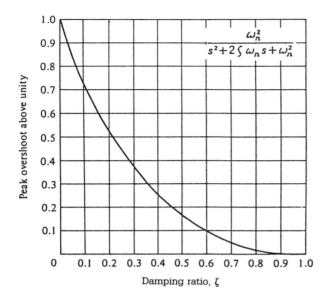

Figure 4.32 Peak overshoot for a damped oscillator.

The closed loop transfer function with unity feedback is of the form

$$\frac{\theta_o(s)}{\theta_i(s)} = \frac{KG(s)}{1 + KG(s)}$$

$$= \frac{P_1(s)}{P_1(s) + P_2(s)}. \qquad (4.82)$$

The root-locus method is a graphical technique for determining the zeros of $P_1(s) + P_2(s)$ from the zeros of $P_1(s)$ and $P_2(s)$, individually. If a system parameter is varied, the corresponding changes in the zeros of $P_1(s)$ and $P_2(s)$ are determined (just as would be done if the gain and phase plots were being used for design), and the resulting changes in the zeros of $P_1(s) + P_2(s)$ are investigated.

The root-locus is the locus of those values of s for which the phase of the open loop transfer function $KG(s)$ is $\pm n\pi$, where n is an odd positive integer.

The root-locus is easy to sketch since multiplication and division of factors $(s - r_i)$, $(s - a_i)$ results in addition and subtraction of their phase angles.

appropriate sets determined, the designer must then plot more accurate loci in order to select one set, to adjust system parameters, and to evaluate the effects of changes in these parameters.

Based on the approach of reference 5, the angle of $G(s)$, written as $\underline{/G(s)}$, at any specific point in the s-plane is conveniently measured in terms of the angles contributed

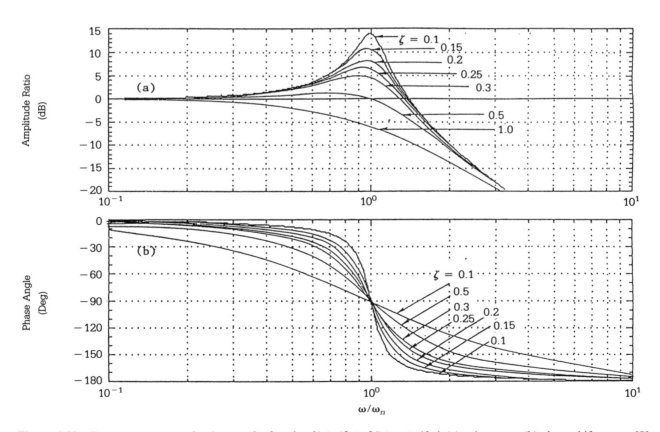

Figure 4.33 Frequency response for the transfer function $[(s/\omega_n)^2 + 2\zeta s/\omega_n + 1]^{-1}$ (a) gain curves (b) phase-shift curves [8].

by the various poles and zeros. In factored form, a typical $G(s)$ for a feedback control system is

$$G(s) = \frac{K(s + z_1)(s + z_2)}{s(s + p_2)(s + p_3)(s + p_4)} \qquad (4.83)$$

which, at the point s_1, becomes

$$G(s_1) = \frac{K(s_1 + z_1)(s_1 + z_2)}{s_1(s_1 + p_2)(s_1 + p_3)(s_1 + p_4)} . \qquad (4.84)$$

The value of $G(s_1)$ can be expressed in terms of the vectors shown in Figure 4.34 as

$$G(s_1) = K \frac{AB}{CDEF} . \qquad (4.85)$$

The angle of $G(s_1)$ is simply determined by the sum and differences of the following vector angles:

$$\underline{/G(s_1)} = \underline{/A} + \underline{/B} - \underline{/C} - \underline{/D} - \underline{/E} - \underline{/F} . \qquad (4.86)$$

Thus, construction of the root loci involves the determination of those points, s_i, in the s plane at which

$$\sum \underline{/\text{vectors from zeros to } s_i}$$
$$- \sum \underline{/\text{vectors from poles to } s_i} = 180° + n360° .$$
$$\qquad (4.87)$$

The construction of the root loci does not, however, entail an aimless search for s-plane points satisfying equation

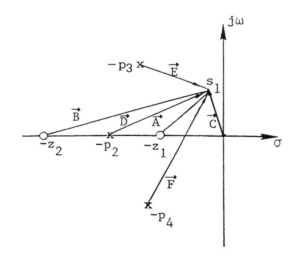

Figure 4.34 Graphical interpretation of $G(s_1)$.

4.87. The root loci are continuous curves, since the zeros of a polynomial are continuous functions of the coefficients. Consequently, once sections of the loci are established, the extension of these parts can be effected without lengthy trial and error. The following observations can aid in constructing the root loci:

1. $K \rightarrow 0$: If $G(s)$ is written as $KG_1(s)$, the equation satisfied along the loci is

$$KG_1(s) = -1 . \qquad (4.88)$$

Clearly, if K tends to zero, $G_1(s)$ must tend to infinity, or s must approach the poles of $G(s)$. Hence, if the loci are interpreted as plots of the closed loop pole positions when the open loop gain varies from 0 to $+\infty$, the loci start at the open loop poles.

2. $K \rightarrow \infty$: Since equation 4.88 indicates that a very large K requires a $G_1(s)$ tending to zero, or a value of s approaching a zero of $G(s)$, the loci terminate on the zeros of $G(s)$.

3. *Number of loci*: From the fact that a polynomial of degree n possesses n zeros, it is clear that the number of separate loci equals the number of poles or zeros of $G(s)$.

4. *Conjugate values*: Complex parts of the loci always appear in conjugate complex pairs if, as is customary in the design of feedback systems, the coefficients of the polynomials $P_1(s)$ and $P_2(s)$ are real.

5. *Loci near infinity*: The behavior of the loci for large values of s is readily investigated by replacing $P_1(s)$ and $P_2(s)$ by the highest powers of each.

As an example, consider a single-loop motor-driven servo in Figure 4.35.

$$KG(s) = \frac{KK_m}{s\left(s + \dfrac{1}{\tau_1}\right)\left(s + \dfrac{1}{\tau_2}\right)} . \qquad (4.89)$$

There are no zeros, but poles exist at 0, $-1/\tau_1$, and $-1/\tau_2$. For any s at point Q, vector s has phase θ and vector $s + 1/\tau_1$ is P_1Q with phase ϕ_1; vector $s + 1/\tau_2$ is P_2Q with phase ϕ_2. The condition that s (point Q) lies on the locus of points for which the phase of $KG(s)$ is $\pm n\pi$ is that $-\theta - \phi_1 - \phi_2 = \pm n\pi$. This condition gives the root-locus plot shown in Figure 4.36. The thick black lines are the root-loci.

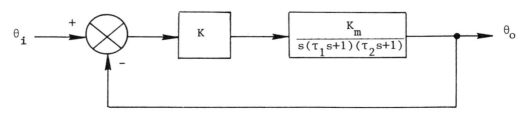

Figure 4.35 Single-loop motor-driven servo.

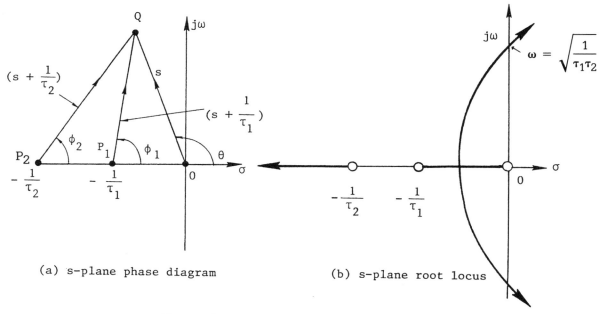

Figure 4.36 Root locus plot for a motor-driven servo.

Next, the zeros of $1 + KG(s)$ are located on the root-locus plot by trial and error. A point s on the root-locus is chosen, $KG(s)$ is obtained by dividing K by the product of the lengths s, $s + 1/\tau_1$, $s + 1/\tau_2$. Repeat with other points on the root-locus until the value 1 is obtained for $KG(s)$. The values of s thus found are the roots of $1 + KG(s) = 0$ and the poles of the closed loop transfer function. Of course, for stability, all these roots will lie in the left half plane.

Note that once all the roots of $1 + KG(s)$ have been determined it becomes possible to expand

$$\Theta_o(s) = \frac{\Theta_i(s)KG(s)}{1 + KG(s)} \qquad (4.90)$$

in partial fractions and thus to obtain the inverse Laplace transform $\theta_o(t)$. A root-locus plot is useful in checking the effect on stability of changing the gain. To illustrate this, consider the same root-locus as before. If $s = a$ is a root of $KG(s) + 1 = 0$, then

$$K = \frac{1}{|G(a)|} = |\overline{Oa}| \cdot |\overline{P_1a}| \cdot |\overline{P_2a}|. \qquad (4.91)$$

If the gain K is increased, the lengths $\overline{Oa}$, $\overline{P_1a}$, and $\overline{P_2a}$ must increase and the root a moves closer to the instability condition. Conversely, a decrease in K produces root movement away from instability. The $j\omega$ axis in the root-locus plot thus corresponds to the $-1 + jO$ point on the Nyquist diagram.

The root-locus, Nyquist, Bode, and Nichols plots for the single-loop motor-driven servo are illustrated in Figure 4.37 through 4.40, respectively.

4.14 Structural Resonance Considerations

The denominating parameter which determines the maximum control response is dependent upon the value of the lowest structural resonance frequency. The open loop gain crossover frequency of the first inner loop (rate loop) should, in general, be about one-third of the structural resonant frequency. If an outer position loop is also used, then the open loop crossover frequency should be typically one-half of the rate loop crossover frequency or one-sixth of the lowest structural resonance frequency. Good design practice occurs when the ratio is between one-eighth and above.

Under large overload conditions, the system should degrade gracefully. All sensors should operate within limits, and under no conditions should sensor outputs vanish, which can result in instabilities.

4.15 Recommended Practice for Active Control Systems

The following recommendations apply to the design of active attitude control systems. Specific recommendations with regard to the thrusting maneuvers and spacecraft with structural flexibility are included.

4.15.1 General Considerations

1. Ensure that all closed loop control systems exhibit acceptable transient response.

2. Control system torque capability must be sufficiently large to correct initial condition errors and maintain

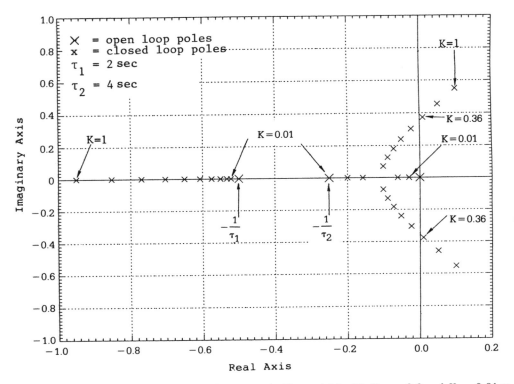

Figure 4.37   Root locus plot for the single-loop motor-driven servo in Figure 4.34 with $Km = 0.3$ and $K = 0.01$ to 1 variation [8].

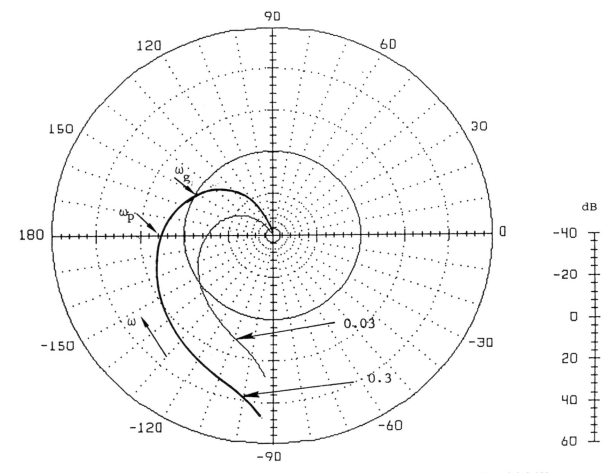

Figure 4.38   Nyquist plot for a motor-driven servo in Figure 4.35 with $KKm = 0.03$ and 0.3 [8].

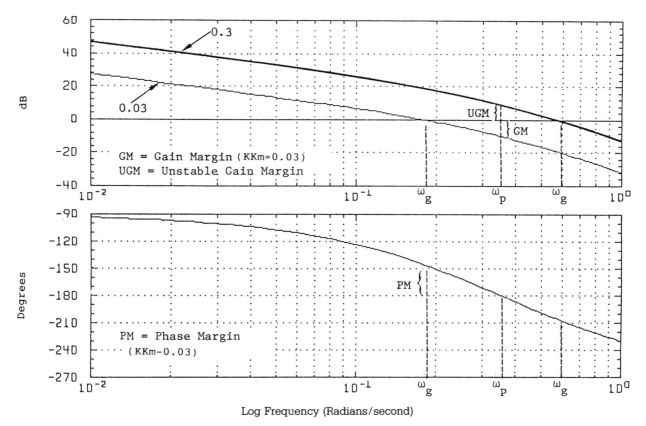

Figure 4.39   Bode plot for a motor-driven servo in Figure 4.35 with *KKm* = 0.03 and 0.3 [8].

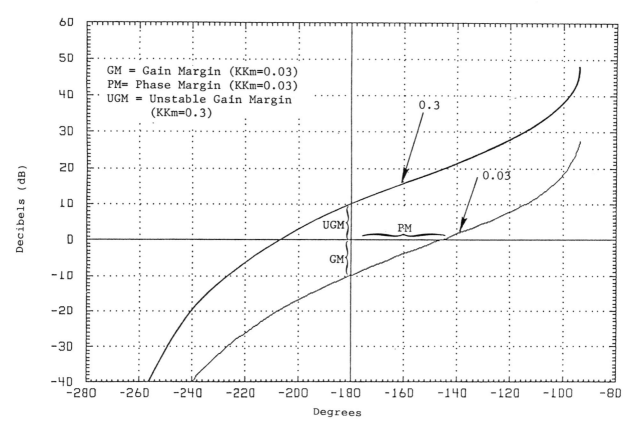

Figure 4.40   Nichols plot for a motor-driven servo in Figure 4.35 with *KKm* = 0.03 and 0.3 [8].

attitude limits within specified values in the presence of the maximum environmental disturbances.

3. The control logic must be consistent with the minimum impulse size and lifetime specification of the thrusters.

4. Evaluate system performance incorporating as many hardware elements in a simulation as possible.

5. Combine the normal tolerances statistically with the beginning and end-of-life center of mass location and moment of inertia characteristics.

4.15.2 Thrusting Maneuvers

1. Maximize the distance between the center of mass and the point of application of the control force.

2. Determine that propellant expenditure is consistent with system constraints and performance. Determine performance in the event of thruster failure.

3. Design limit cycling characteristics to avoid excessive error, excitation of the spacecraft flexible modes, excessive propellant expenditure, and excessive thruster wear.

4. Consider the effect of thrust impingement on thermal design and thrust degradation.

5. Keep the residual conditions at end of maneuver within the control capability of the control system to prevent excessive expenditure of propellants.

6. Provide capability to verify control system operation before thrust initiation.

7. Maintain current and accurate mass properties of the spacecraft and alternate configurations.

8. Consider all sources of thrust misalignment, including:

a. Thrust vector to thruster misalignment.
b. Thruster mechanical misalignment.
c. Thruster support structure compliance.

4.15.3 Structural Flexibility

1. Provide adequate separation of the rigid body and flexible mode frequencies.

2. Choose the control bandpass so that it excludes the structural resonant frequencies by stiffening of the structure. Consider the use of special inner control loops, or the use of notch filters.

3. Make the damping ratio of each flexible appendage as large as possible. Artificial stiffening or damping by means of separate control loops should be considered.

4. Under saturation, all loops should degrade gracefully.

4.16   References

1. Humphrey, W. M. *Introduction to Servomechanism System Design*, Prentice-Hall, 1973.
2. James, H. M., N. B. Nichols, and R. S. Phillips. *Theory of Servomechanics*, McGraw-Hill, 1947.
3. Kane, T. R. *Dynamics*, 2nd ed., Stanford University, 1972.
4. Palm, W. J. *Control Systems Engineering*, John Wiley and Sons, 1986.
5. Truxal, J. G. *Automatic Feedback Control System Synthesis*, McGraw-Hill, 1955.
6. Chang, S. S. L. *Fundamental Handbook of Electrical and Computer Engineering*, R. Gram Editor, Vol. II, Section 4, Wiley Interscience, 1983.
7. Takahashi, Y., M. J. Rabins, and D. M. Auslander. *Control and Dynamic Systems*, Addison-Wesley, 1970.
8. Lee, E. A. *Control Analysis Program for Linear Systems (CAPLIN)*, Version 1.02, The Aerospace Corporation, December 1989.
9. Martin, D. H. *Communication Satellites 1958 to 1982*, The Aerospace Corporation, Report No. TR-79–078, 31 December 1986.

Chapter 5
Momentum Exchange Systems

5.1 Introduction

Attitude control systems containing reaction wheels (RWs) or control moment gyros (CMGs) are known as momentum exchange systems. The angular momentum absorbed by such devices can be transferred to the spacecraft or satellite.

The main advantage of the momentum exchange system over a pure mass expulsion system is a more efficient capability for fine attitude control. Also, if a significant number of slewing maneuvers are required, the use of a momentum exchange system may save more propellant mass than the additional mass required to have exchange actuators (which consume power rather than mass).

Control moment gyros run at constant speed and may have one or two gimbals, depending on the degrees of freedom. They absorb angular momentum by gimbal rotation. The resultant angular momentum of a system of constant magnitude vectors is changed by altering their direction. Reaction wheels, on the other hand, absorb angular momentum by accelerating and decelerating an inertia wheel fixed in the satellite. The magnitude of the angular momentum vector is thus changed rather than its direction.

The principal advantage of CMGs compared to RWs is their large torque capability with linearity at low power, as well as lower weight, power, and size requirements for the same performance capability. The main disadvantage is a somewhat greater complexity and a computer requirement. The use of the RWs is therefore indicated where relatively low torque and angular momentum storage capacity is required and where no complex maneuvering operations are necessary. Reaction wheels operating at constant angular velocities which can be modulated (varied) are known as momentum bias systems. Such systems offer good resistance to external disturbances and fine pointing control. An example of such a system will be presented.

Two degree of freedom CMGs generally have size, weight, and power advantages over single degree of freedom (single gimbal) gyros. They have much larger torquers, however, since the relatively high output torques are not directly transmitted to the vehicle bearings but must be reacted through the torquers.

Excellent system response, however, can be achieved by using single degree of freedom gyros in twin pairs. The inter-axis coupling is thus minimized and the performance can be improved to the point where a choice between the single or two degree of freedom gyros is not obvious.

5.2 Reaction Wheel Systems

Consider a single axis system such as is shown in Figure 5.1 [1]. Let I and J be the moments of inertia of the vehicle and wheel, respectively.

The magnitude of the angular momentum of the wheel is

$$H_w = J(\dot{\theta} + \Omega) \qquad (5.1)$$

where

Ω = angular rate of the wheel relative to the body (vehicle)
$\dot{\theta}$ = angular rate of the body

The magnitude of the angular momentum of the vehicle is

$$H_v = I\dot{\theta} \; . \qquad (5.2)$$

For a single degree of freedom planar motion of both the vehicle and the wheel, the equation of motion is of the form

$$\frac{d}{dt}(H_w + H_v) = T \qquad (5.3)$$

or

$$(I + J)\ddot{\theta} + J\dot{\Omega} = T \qquad (5.4)$$

where T is the external disturbance torque value. Taking the Laplace transformation of both sides of this equation yields

$$(I + J)s^2\bar{\theta} + Js\bar{\Omega} = \bar{T} \qquad (5.5)$$

where bars denote the respective transformed quantities. For the vehicle only

$$Is^2\bar{\theta} = \bar{T} + \bar{T}_c \qquad (5.6)$$

and for the wheel only, one subtracts equation 5.6 from equation 5.5 to get

$$Js(\bar{\Omega} + s\bar{\theta}) = -\bar{T}_c \qquad (5.7)$$

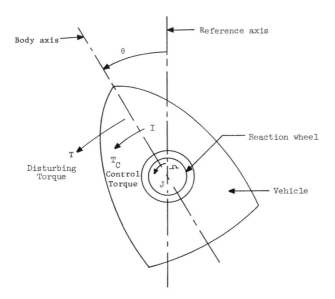

Figure 5.1 Vehicle and wheel system.

where $\overline{T}_C$ is the wheel control torque. Friction torque and motor back electromotive force (emf) are assumed to be zero. A schematic diagram for this system is shown in Figure 5.2.

The upper part in Figure 5.2 represents equation 5.6. To this is added a simple control loop employing proportional (K) plus derivative (Ks/σ) control. Let the gain $K = I/\tau^2$, where τ is the time constant of the system. Thus, consider the above as part of a general feedback control system as shown in Figure 5.3, where

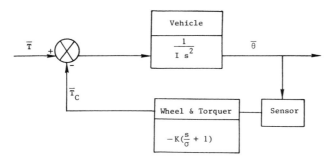

Figure 5.2 Reaction wheel control system.

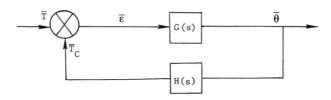

Figure 5.3 General feedback system.

$$G(s) = \frac{1}{Is^2} \qquad (5.8)$$

and

$$H(s) = K\left(\frac{s}{\sigma} + 1\right)$$
$$= K_R s + K_A . \qquad (5.9)$$

Here K_R and K_A are the rate and position gains. The open loop transfer function is of the form

$$\frac{\overline{T}_C}{\overline{\varepsilon}} = \frac{\overline{T}_C}{\overline{T} - \overline{T}_C} = G(s)\,H(s) \qquad (5.10)$$

and the closed loop transfer function takes the form

$$\frac{\overline{\theta}}{\overline{T}} = \frac{G}{1 + G(s)\,H(s)} = \frac{1}{Is^2 + K\left(\dfrac{s}{\sigma} + 1\right)} . \qquad (5.11)$$

The poles (roots) of this equation are

$$S_{1,\,2} = \frac{-K/\sigma + \sqrt{(K/\sigma)^2 - 4KI}}{2I} . \qquad (5.12)$$

If $(K/\sigma)^2 = 4KI$ (critical damping), then $S_{1,\,2} = -2\sigma$; that is, the two roots are negative and equal. If we now define a time constant $\tau = 1/2\sigma$, where $\tau^2 = I/K$, then

$$\frac{\overline{\theta}}{\overline{T}} = \frac{1}{K\left(\dfrac{I}{K}s^2 + \dfrac{s}{\sigma} + 1\right)}$$
$$= \frac{1}{K(\tau^2 s^2 + 2\tau s + 1)} = \frac{\tau^2}{I(\tau s + 1)^2} . \qquad (5.13)$$

5.2.1 Response to Attitude Error

The response of this system to an initial attitude error $\theta(0)$ is of the form

$$\theta = \theta(0)\left(1 + \frac{t}{\tau}\right)e^{-t/\tau} \qquad (5.14)$$

from which the required constant τ can be computed for a desired speed of response.

5.2.2 Response to an Impulse

For a torque impulse of magnitude i, the system response is

$$\theta(t) = \frac{i\tau}{I}\frac{t}{\tau}e^{-t/\tau} \qquad (5.15)$$

which is shown in Figure 5.4.
Maximum system response occurs at $t = \tau$ and is given by equation 5.15 as

$$\theta_{max} = \frac{i\tau}{I} \cdot \frac{1}{e} . \qquad (5.16)$$

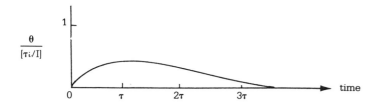

Figure 5.4 Response to an impulse torque of magnitude i.

5.2.3 Control Torque Required

From Figure 5.2, the control torque is

$$\overline{T}_C = -K\left(\frac{s}{\sigma} + 1\right)\overline{\theta} \ . \qquad (5.17)$$

Substituting for $\overline{\theta}$ from equation 5.13 and simplifying yields the following:

$$\overline{T}_C = -K\left(\frac{s}{\sigma} + 1\right)\frac{\tau^2 \overline{T}}{I(\tau s + 1)^2}$$

$$= -\frac{I}{\tau^2}(2\tau s + 1)\frac{\tau^2 \overline{T}}{I(\tau s + 1)^2}$$

$$= -\frac{(2\tau s + 1)\overline{T}}{(\tau s + 1)^2} \ . \qquad (5.18)$$

For a torque impulse of magnitude i, the control torque will be of the form

$$\overline{T}_C(t) = -\frac{2i}{\tau}\left(1 - \frac{t}{2\tau}\right)e^{-t/\tau} \qquad (5.19)$$

which is divided by i/τ and plotted as a function of time in Figure 5.5.

The peak (negative) value of control torque $\overline{T}_C$ occurs at $t = 0$ and is given as

$$\overline{T}_{C_{max}} = -\frac{2i}{\tau} \ . \qquad (5.20)$$

5.3 Momentum Bias System

Consider a typical geosynchronous orbital spacecraft as shown in Figure 5.6. A single RW is aligned along the pitch

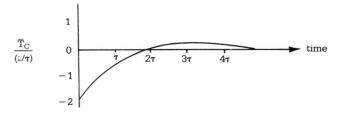

Figure 5.5 The time dependence of the control torque $\overline{T}_C$ divided by i/τ.

axis of the spacecraft which is oriented along the normal to the orbital plane.

In addition to the reaction control (thruster) subsystem, the satellite also carries an attitude and velocity control subsystem (AVCS) package which provides for attitude error sensing, data storage, signal processing, mode switching, verification of commands, and control of the actuation devices. The AVCS operates in a spin stabilized mode during the transfer orbit; provides for spin speed adjustment, despin, and Sun/Earth acquisition control; and operates in a three axis on-orbit control mode. A relatively small body-fixed momentum (reaction control) wheel is used in conjunction with low level (0.5 N) hydrazine thrusters and an Earth sensor to provide accurate three-axis on-orbit control utilizing only roll and pitch attitude sensing. The AVCS also provides for ground commanded orbital velocity changes while in a three axis thruster control mode.

The reaction wheel assembly (RWA) is an inertia wheel that controls the pointing of the spacecraft about the pitch or Y axis during the normal mode of operation. The RW spins at a nominal bias speed of 3,000 revolutions per minute. The housing of the RWA is hard mounted to the spacecraft with the spin axis of the inertia wheel oriented parallel to the spacecraft pitch axis. Acceleration or deceleration of the RW produces corrective torques on the spacecraft to maintain the pointing accuracy of the spacecraft to the center of the Earth to within a small angle, e.g., 0.25 degree. Gyroscopic coupling of the RWA bias momentum with the spacecraft orbital rate provides passive control of the spacecraft about the yaw or Z axis. A model reaction wheel assembly is illustrated in Figure 5.7.

During spacecraft operations, the RW is brought up to cage speed and maintained there until acquisition is complete. (Caging the RW occurs when the wheel is set at a constant speed, in this case, 3,000 revolutions per minute, and the wheel no longer controls pitch). After acquisition, the RW functions as a portion of a control loop including the Earth sensors. Depending on signals from these sensors, the wheel will be accelerated or decelerated providing a reaction torque to the spacecraft in a direction to reduce the Earth pointing error. When the RW speed varies from the bias speed by more than 500 revolutions per minute, a cage command is given which drives the RW to cage speed. At the same time, reaction control system thruster jets are fired to compensate for the reaction torques applied to the spacecraft by the RW during the caging operation. The RW is also commanded to go to cage speed whenever velocity correction thruster firings are initiated.

The spacecraft pitch axis (RW spin axis) is nominally parallel to the axis of the spacecraft's orbit about the Earth. If the spacecraft rotates away from its nominal position about its yaw axis, then the bias momentum of the RW will couple with the spacecraft's orbit rate providing gyroscopic

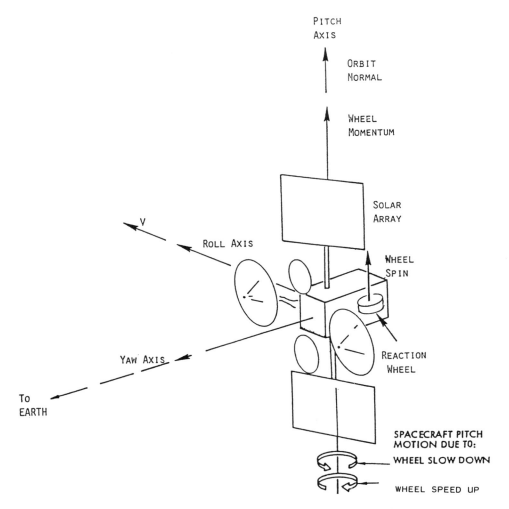

Figure 5.6 Momentum bias system (variant of dual-spin).

torques to return the spacecraft to its proper orientation about the yaw axis.

Long term on-orbit roll/yaw control is implemented using the momentum bias afforded by the speed biased pitch wheel, operated in concert with low level offset roll/yaw thrusters. Roll/yaw momentum accumulation and nutation damping are obtained by operating the offset thrusters on the basis of roll attitude error data from the Earth sensors. These thrusters are canted (''offset'') in the roll/yaw plane so that a roll control torque is accompanied by a yaw torque. This offset feature provides damping of orbital frequency motions. An example of a spacecraft controlled by a biased momentum RW is illustrated in Figure 5.8.

5.3.1 Equations of Motion

Consider a satellite in a circular orbit with the reference axes as shown in Figure 5.9.

The satellite principal body axes e_α can be related to the orbital axes $E_\beta(\alpha, \beta = 1, 2, 3)$ as

$$\begin{pmatrix} e_1 \\ e_2 \\ e_3 \end{pmatrix} = R(\psi)\, R(\theta)\, R(\phi) \begin{pmatrix} E_1 \\ E_2 \\ E_3 \end{pmatrix} \tag{5.21}$$

where the rotation matrices are

$$R(\phi) = \begin{pmatrix} 1 & 0 & 0 \\ 0 & c\phi & s\phi \\ 0 & -s\phi & c\phi \end{pmatrix},\ R(\theta) = \begin{pmatrix} c\theta & 0 & -s\theta \\ 0 & 1 & 0 \\ s\theta & 0 & c\theta \end{pmatrix}$$

$$R(\psi) = \begin{pmatrix} c\psi & s\psi & 0 \\ -s\psi & c\psi & 0 \\ 0 & 0 & 1 \end{pmatrix}.$$

The angular velocity components ω_1, ω_2, and ω_3 in the spacecraft body coordinates are given by

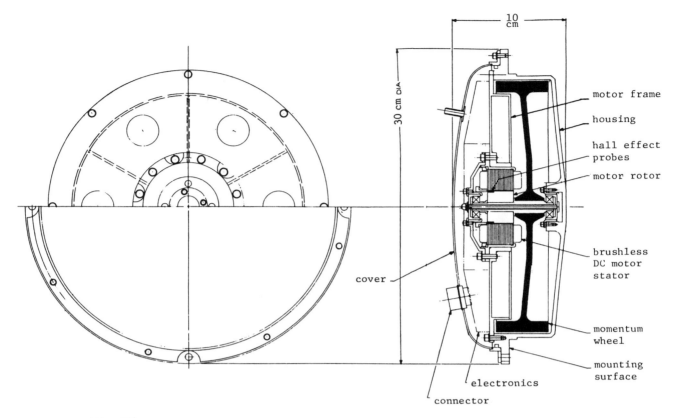

W = 7 kg
NORMAL RPM = 2,000
WHEEL INERTIA = 0.03 kg · m^2
POWER = 1-24 watts (peak)

Figure 5.7   Model reaction wheel assembly.

$$\begin{pmatrix} \omega_1 \\ \omega_2 \\ \omega_3 \end{pmatrix} = \begin{pmatrix} 0 \\ 0 \\ \dot{\psi} \end{pmatrix} + R(\psi) \begin{pmatrix} 0 \\ \dot{\theta} \\ 0 \end{pmatrix} + R(\psi)R(\theta) \begin{pmatrix} \dot{\phi} \\ 0 \\ 0 \end{pmatrix}$$

$$+ R(\psi)\, R(\theta)\, R(\phi) \begin{pmatrix} 0 \\ \omega_o \\ 0 \end{pmatrix}. \quad (5.22)$$

Neglecting nonlinear terms and considering only small angles yields

$$\begin{pmatrix} \omega_1 \\ \omega_2 \\ \omega_3 \end{pmatrix} \approx \begin{pmatrix} \dot{\phi} \\ \dot{\theta} \\ \dot{\psi} \end{pmatrix} + R(\psi)\, R(\theta)\, R(\phi) \begin{pmatrix} 0 \\ \omega_o \\ 0 \end{pmatrix}$$

$$\approx \begin{pmatrix} \dot{\phi} \\ \dot{\theta} \\ \dot{\psi} \end{pmatrix} + \begin{pmatrix} \omega_o\psi \\ \omega_o \\ -\omega_o\phi \end{pmatrix}. \quad (5.23)$$

In general, the total angular momentum is

$$\vec{H} = I_1\omega_1\hat{e}_1 + I_2\omega_2\hat{e}_2 + I_3\omega_3\hat{e}_3 + h_2\hat{e}_2 \quad (5.24)$$

where h_2 is the RW angular momentum along the e_2 axis relative to the body fixed frame e_α. In the body frame, the external torque equation is

$$\dot{\vec{H}} + \vec{\omega} \times \vec{H} = \vec{T}. \quad (5.25)$$

The linearized component equations for the roll, pitch, and yaw motion become

$$I_1\ddot{\phi} + [\omega_o^2(I_2 - I_3) + \omega_o h_2]\phi$$
$$+ [\omega_o(I_1 - I_2 + I_3) - h_2]\dot{\psi} = T_1$$
$$I_2\ddot{\theta} = T_2$$
$$I_3\ddot{\psi} + [\omega_o^2(I_2 - I_1) + \omega_o h_2]\psi$$
$$- [\omega_o(I_1 - I_2 + I_3) - h_2]\dot{\phi} = T_3. \quad (5.26)$$

With torque components taken as zero, the characteristic equation for the system, obtained by taking the Laplace transform, is of the form

$$(s^2 + \omega_o^2)(s^2 + \omega_1^2) = s^4 + s^2(\omega_o^2 + \omega_1^2) + \omega_o^2\omega_1^2 \quad (5.27)$$
$$= 0$$

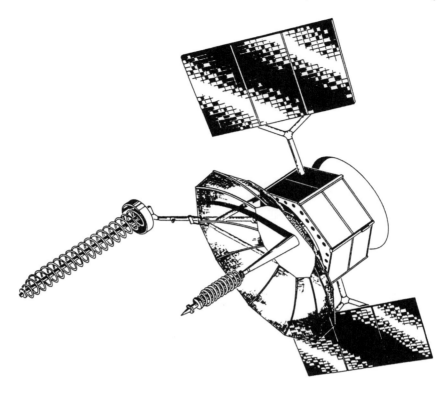

FLTSATCOM satellite.

Figure 5.8 A spacecraft controlled by a biased momentum RW.

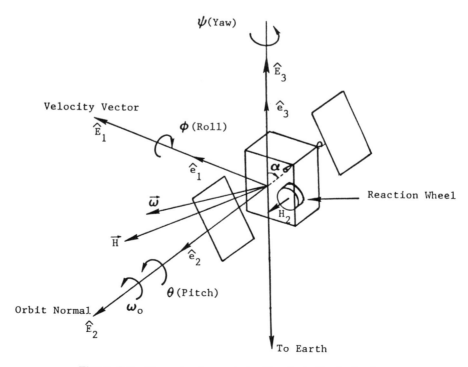

Figure 5.9 Geometry for control with a body fixed wheel.

where

$$\omega_1^2 = \frac{h_2^2}{I_1 I_3} + \omega_o \left(\frac{2 I_2 h_2}{I_1 I_3} - \frac{h_2}{I_3} - \frac{h_2}{I_1} \right)$$
$$+ \omega_o^2 \left(1 + \frac{I_2^2}{I_1 I_3} - \frac{I_2}{I_3} - \frac{I_2}{I_1} \right) . \quad (5.28)$$

This system, therefore, always has a root at orbital frequency $\omega_o (s = \pm j\omega_o)$, undergoing nutational motion at angular frequency ω_1, where

$$\omega_1 \approx \frac{h_2}{\sqrt{I_1 I_3}} . \quad (5.29)$$

In general, $\omega_1 \gg \omega_o$.

In equation 5.26, the T_1, T_2, and T_3 terms are the external torques acting on the spacecraft such as gravity gradient, solar radiation, etc.

5.3.2   Roll-Yaw Control

Equation 5.26 shows that the roll and yaw motions (ϕ, ψ) are uncoupled from the pitch motion (θ). The linearized equations with small momentum terms deleted are

$$I_1 \ddot{\phi} - h_2 \dot{\psi} + \omega_o h_2 \phi = T \cos \alpha + T_{s1}$$
$$I_3 \ddot{\psi} + h_2 \dot{\phi} + \omega_o h_2 \psi = T \sin \alpha + T_{s3} \quad (5.30)$$

where T is the magnitude of the torque delivered by the offset thrusters, and T_{s1} and T_{s3} are the external torque components (due to solar radiation) acting on the roll and yaw axes, respectively.

Figure 5.10 shows a typical offset thruster control channel, consisting of a noise filter, a derived-rate modulator,

and valve drive electronics. The derived-rate modulator shown is a dual time constant design, exhibiting one short-time constant (τ_1) while "on" and another longer time constant (τ_o) while "off." This arrangement allows the modulator to fulfill its two primary functions (metering out short impulse bits and providing lead information for nutation damping) without unduly constraining the modulator hysteresis $h\phi_D$, where ϕ_D is the roll channel deadband.

The steady state solutions can be obtained from equation 5.30 as

$$\omega_o h_2 \phi_{ss} = T \cos \alpha + T_{s1}$$
$$\omega_o h_2 \psi_{ss} = T \sin \alpha + T_{s3} \quad (5.31)$$

where

$$T = -K(\tau_1 \dot{\phi} + \phi) \quad \text{and} \quad K = 1/K_f \text{ in Figure 5.10.}$$

Under steady state conditions

$$T_{ss} = -K\phi_{ss} . \quad (5.32)$$

Therefore

$$\phi_{ss} = \frac{T_{s1}}{\omega_o h_2 + K \cos \alpha} \quad (5.33)$$

and

$$\psi_{ss} = \frac{T_{s3} - K\phi_{ss} \sin \alpha}{\omega_o h_2} . \quad (5.34)$$

It can be seen that ϕ_{ss} and ψ_{ss} are inversely proportional to wheel momentum h_2.

For zero roll error $\phi_{ss} = 0$ and $T_{s1} = 0$. The yaw angle dynamics are determined from equation 5.30, which becomes

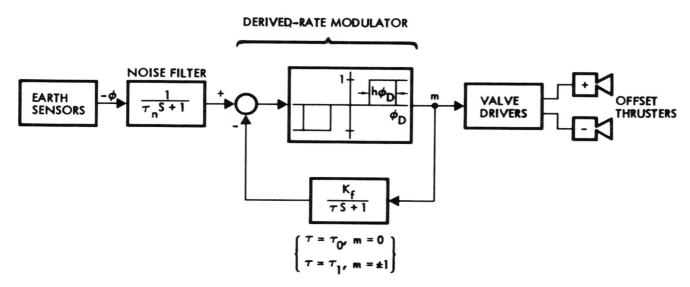

Figure 5.10   A typical offset thruster controller.

$$h_2 \dot{\psi} = -T \cos \alpha$$

$$I_3 \ddot{\psi} + \omega_o h_2 \psi = T \sin \alpha + T_{s3}$$

$$= -\left(\frac{h_2 \dot{\psi}}{\cos \alpha}\right) \sin \alpha + T_{s3} \quad (5.35)$$

Therefore,

$$I_3 \ddot{\psi} + h_2 \dot{\psi} \tan \alpha + \omega_o h_2 \psi = T_{s3} \quad (5.36)$$

This is a damped second order differential equation with a steady state solution

$$\psi_{ss} = \frac{T_{s3}}{\omega_o h_2} . \quad (5.37)$$

Thus, ψ_{ss} is directly proportional to the steady state yaw disturbance torque and inversely proportional to wheel momentum.

Equation 5.36 shows that h_2 and α determine the damping characteristics of the yaw (ψ) motion as well as the roll (ϕ) motion through coupling in terms of equation 5.30.

In general, however, the linear control model of equation 5.30 will not provide a realistic picture of system dynamics because it ignores the deadband of the modulator. During convergence motion from outside the deadband, the motion can indeed be studied using equations 5.30 and 5.31 in combination. Motion inside the deadband is characterized (in the absence of disturbance torques) by nutational motion, which is virtually undamped, superimposed upon orbit frequency excursions providing roll/yaw momentum exchange. With forcing from environmental disturbances, the roll error will move between the positive and negative deadbands at orbit frequency, with modulator induced thruster activity occurring for a portion of each half orbit. The long term roll and yaw motion characteristics are illustrated in Figure 5.11.

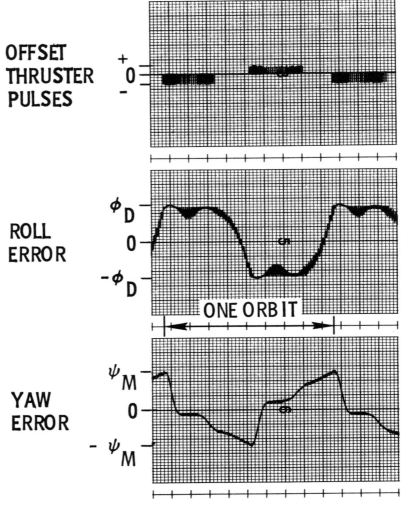

Figure 5.11 Long term roll and yaw motion characteristics.

5.3.3 Pitch On-Orbit Wheel Control

Pitch axis control is independent (uncoupled) from roll/yaw control and is accomplished by modulating the wheel speed above and below the nominal spin speed. The general dynamical characteristics are described by equations 5.1 through 5.20. A typical wheel control block diagram is shown in Figure 5.12. The integral and lead compensation blocks serve to drive the pitch error to zero and ensure good transient response with sufficient damping. The limiter prevents the wheel speed from exceeding design limits, and Earth sensors provide the pitch error signal. The RW provides fine control about the pitch axis (typically 0.05 degree or less). Momentum unloading may occur at 10 percent speed variation.

Lead compensation provides the correct phase in the control loop for proper damping. An ordinary lead network, however, acts like a high pass filter, causing high frequency noise components of a signal to appear to be greatly amplified. An example of a frequency response plot for a lead compensation transfer function with a double pole at 1.50 radians per second is illustrated in Figure 5.13. This has the effect of reducing the gain at high frequencies.

5.4 Control Moment Gyro Systems

The equations of motion for a single degree of freedom gyro are derived and analyzed in the following section, in order to present the main characteristics of these systems.

5.4.1 Single CMG Control System Dynamics

A single gimbal control moment gyro is shown schematically in Figure 5.14 and pictorially in Figure 5.15.

Following the approach of reference 2, the reaction torque exerted by the CMG rotor on the gimbal, which is attached to the spacecraft along the e_2 body axis, can be expressed as

$$\vec{T}_G = -\frac{d\vec{H}}{dt} = -\dot{\vec{H}} - \vec{\omega} \times \vec{H} \qquad (5.38)$$

where $\vec{H}$, the CMG angular momentum, is

$$\vec{H} = \mathrm{H} \sin \delta \hat{e}_1 + H \cos \delta \hat{e}_3 \qquad (5.39)$$

and

$$\vec{\omega} = \omega_1 \hat{e}_1 + \omega_2 \hat{e}_2 + \omega_3 \hat{e}_3 \qquad (5.40)$$

is the angular velocity of the spacecraft. $\hat{e}_1$, $\hat{e}_2$, and $\hat{e}_3$ are unit vectors along the principal axes of the vehicle, and δ is the precession angle of the gimbal. The gimbal also receives torques T_m and $D\dot{\delta}$ from the torque motor and damper, respectively. The total torque on the gimbal is then

$$\begin{aligned}
\vec{T}_G = &-H(\omega_2 + \dot{\delta})\cos \delta \hat{e}_1 \\
&+ [T_m - D\dot{\delta} + H(\omega_1 \cos \delta - \omega_3 \sin \delta)]\hat{e}_2 \\
&+ H(\omega_2 + \dot{\delta})\sin \delta \hat{e}_3 .
\end{aligned} \qquad (5.41)$$

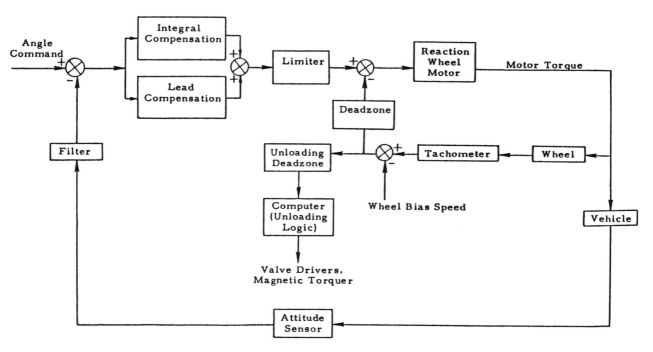

Figure 5.12 Pitch wheel (momentum bias) attitude control block diagram.

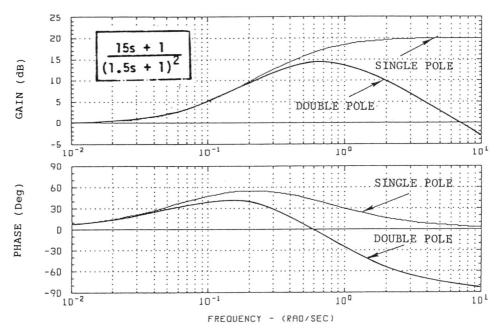

Figure 5.13 Frequency response plot for a lead compensation transfer function with a double pole at 1.50 radians per second.

5.4.2 Component Equations

The e_2 torque component from equation 5.41 yields the gimbal equation of motion

$$T_m - D\dot{\delta} + H(\omega_1 \cos \delta - \omega_3 \sin \delta)$$
$$= I_G(\ddot{\delta} + \dot{\omega}_2) \quad (5.42)$$

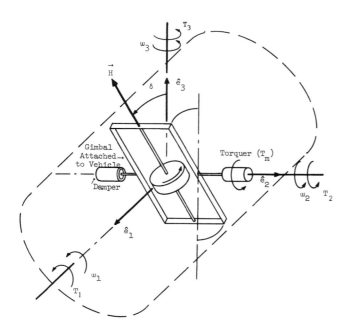

Figure 5.14 A single degree of freedom CMG system.

where I_G is the combined rotor and gimbal moment of inertia about the gimbal axis.

The e_1 and e_3 torque components are transmitted directly to the vehicle by the gimbal bearings. The vehicle component equations of motion are

$$I_1\dot{\omega}_1 - (I_2 - I_3)\omega_2\omega_3 = T_1 - H(\omega_2 + \dot{\delta})\cos \delta$$
$$I_2\dot{\omega}_2 - (I_3 - I_1)\omega_3\omega_1 = T_2 - T_m + D\dot{\delta}$$
$$I_3\dot{\omega}_3 - (I_1 - I_2)\omega_1\omega_2 = T_3 + H(\omega_2 + \dot{\delta})\sin \delta \quad (5.43)$$

where T_1, T_2, and T_3 are external torque magnitudes, I_1, I_2, I_3 and ω_1, ω_2, ω_3 are the vehicle yaw, pitch, and roll moments of inertia and angular velocities, respectively.

For a small gimbal angle δ, the vehicle motion is essentially restricted to the e_1 axis, and neglecting nonlinear terms (products) the linearized and decoupled equations 5.30 and 5.31 become

$$T_m - D\dot{\delta} + H\omega_1 = I_G\ddot{\delta} \quad (5.44)$$

$$T_1 - H\dot{\delta} = I_1\dot{\omega}_1 \quad (5.45)$$

where it has been assumed that $\omega_2 = \omega_3 = T_2 = T_3 = 0$. An external disturbance T_1 induces a rate $-\omega_1$, which causes the CMG to precess at angular velocity $\dot{\delta}$ in a direction to oppose the original disturbance. If the disturbance persists, the vehicle will continue drifting while δ increases. When δ reaches 90 degrees, the momentum transfer is complete and the CMG is saturated. The stored disturbance impulse can be removed by reaction control jets.

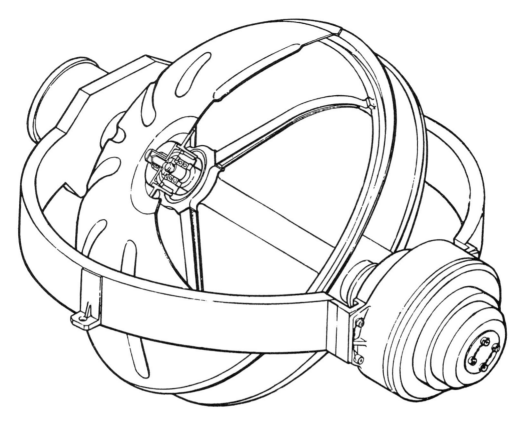

Figure 5.15 A single gimbal CMG system.

An analysis of the linearized equations for the vehicle and gyro can be made as follows. For $T_1 = 0$, equation 5.45 becomes

$$I_1\dot{\omega}_1 + H\dot{\delta} = 0 \qquad (5.46)$$

which can be integrated to yield

$$\omega_1 = -\frac{H\delta}{I_1}. \qquad (5.47)$$

Substituting equation 5.47 into equation 5.44 yields

$$I_G\ddot{\delta} + D\dot{\delta} + \frac{H^2\delta}{I_1} = T_m. \qquad (5.48)$$

This equation shows that the torque T_m (supplied by the torquer) can be very small if $I\ddot{\delta}_G$ is sufficiently small and I_1 is large. The gyroscopic output torque about the e_1 axis, on the other hand, has a magnitude

$$T_o = |\vec{T}_o| = |\dot{\vec{\delta}} \times \vec{H}| = H\dot{\delta} = -I_1\dot{\omega}_1. \qquad (5.49)$$

Therefore, T_o can be very large for large H and $\dot{\delta}$ so that the ratio of input to output torque for steady state ($\ddot{\delta} = 0$) is

$$\frac{T_m}{T_o} = \frac{D\dot{\delta} + (H^2\delta/I_1)}{H\dot{\delta}}$$

$$= \frac{D}{H} + \frac{H\delta}{I_1\dot{\delta}}. \qquad (5.50)$$

This torque ratio can be very small for small D and δ and large I_1 and $\dot{\delta}$. This illustrates the property of torque amplification obtainable from a CMG system having a single degree of freedom.

Some examples of the control systems with reaction wheels and control moment gyros are considered in references 3–6. Certain effects of structural flexibility are examined in references 7–10. Applications involving quaternion representation of orientation and angular velocity are presented in references 11–13. A momentum bias attitude control system is illustrated in reference 14.

5.5 References

1. Cannon, R. H., Jr. "Some Basic Response Relations for Reaction-Wheel Attitude Control," American Rocket Society Meeting in Los Angeles, Preprint 260–11, June 1961.
2. Jacot, A. D., and D. J. Liska. "Control Moment Gyros in Attitude Control," Journal of Spacecraft and Rockets, Vol. 3, No. 9, September 1966.

3. Colburn, B. K., and L. R. White. "Computational Considerations for a Spacecraft Attitude Control System Employing Control Moment Gyros," Journal of Spacecraft and Rockets, Vol. 14, No. 1, January 1977.

4. O'Connor, B. J., and L. A. Morine. "A Description of the CMG and Its Application to Space Vehicle Control," Journal of Spacecraft and Rockets, Vol. 6, No. 3, March 1969.

5. Agrawal, B. N. *Design of Geosynchronous Spacecraft*, Prentice-Hall, Inc., 1986.

6. Wertz, J. R., ed. *Spacecraft Attitude Determination and Control*, D. Reidel Publishing Co., 1980.

7. D'Souza, A. F., and V. K. Gorg. *Advanced Dynamics—Modeling and Analysis*, Prentice-Hall, Inc., 1984.

8. Skelton, R. E. *Dynamic Systems Control*, John Wiley & Sons, 1988.

9. Singh, R. P., Vander Voort, R. J., and P. W. Likins. "Dynamics of Flexible Bodies in Tree Topology—A Computer-Oriented Approach," AIAA Journal of Guidance, Control, and Dynamics, Vol. 8, No. 5, Sept.-Oct. 1985, pp. 584–590.

10. Ho, J. Y. L., and D. R. Herber. "Development of Dynamics and Control Simulation of Large Flexible Space Systems," Journal of Guidance, Control, and Dynamics, Vol. 8, May-June 1985, pp. 347–383.

11. Kane, T. R., Likins, P. W., and D. A. Levinson. *Spacecraft Dynamics*, McGraw-Hill Book Co., New York, 1983.

12. Dwyer, T. A. W. III. "Exact Nonlinear Control of Large Angle Rotational Maneuvers," IEEE Transactions on Automatic Control, Vol. AC-29, No. 9, pp. 769–774, Sept. 1984.

13. Carrington, C. K., and J. L. Junkins. "Nonlinear Feedback Control of Spacecraft Slew Maneuvers," Journal of Astronautical Sciences, Vol. 32, No. 1, Jan.-Mar. 1984, pp. 29–45.

14. Lebsock, K. L. "High Pointing Accuracy With a Momentum Bias Attitude Control System," AIAA Paper No. 78–569, 1978.

Chapter 6

The Environmental Effects

From the standpoint of stability and control, the principal environmental effects which concern a spacecraft designer are solar radiation pressure, gravity gradients, aerodynamics, magnetics, and the meteoroid or man-made debris flux. Other perturbing factors which may be important include the effects associated with internal moving parts, thrust misalignment, thermal emissivity, electromagnetic radiation, outgassing, and propellant leakage. Solar radiation pressure is generally a significant source of attitude and trajectory errors for high altitude (> 1,000 km) and/or interplanetary spacecraft. Gravity gradients, which result from the extended dimensions of the spacecraft, may either cause disturbing or perturbing torques or provide restoring torques when the effect is used for attitude control. Gravity gradients, as well as the magnetic torques caused by the interaction of the spacecraft's magnetic materials with the planetary magnetic field, are most significant at low altitudes (< 1000 km). Similarly, aerodynamic effects are significant only below 500 kilometers altitude and are generally negligible above 1,000 kilometers. Aerodynamic torques acting on the spacecraft are functions of spacecraft geometry and attitude. Because of the highly rarefied atmosphere and large mean free particle paths, the ordinary kinetic theory of gasses is not applicable.

The torques arising from internal moving parts such as the rotating wheels, circulating fluids, scanning devices, etc., must be included in the general equations of motion of the vehicle. Thrust misalignment torques are caused by the thrust line of action not passing through the vehicle's center of mass. Torques from propellant leakage and/or outgassing of spacecraft materials are of a similar nature. Finally, the meteoroid and man-made debris environment must also be considered in certain cases as a possible source of external torques acting on a spacecraft. Collisions with large objects, however, are more likely to result in catastrophic failures rather than possible increases in spacecraft angular rates about some axis. The variation as a function of altitude of the typical disturbance torques acting on a small satellite is illustrated in Figure 6.1.

6.1 Solar Radiation Pressure

The Sun radiates energy with a reasonably well-defined intensity and spectrum. The rate at which the solar energy at all wavelengths is received at the Earth's mean distance from the Sun is known as the solar constant and has a value of $I = 1353 \pm 20 \ W/m^2$.

When radiant energy E falls on a surface, the surface is subject to a force per unit area. The incoming photons may be considered to have an equivalent mass m in accordance with the electromagnetic mass-energy equivalence relationship [1]

$$E = mc^2 \tag{6.1}$$

where c is the velocity of light. The equivalent increment of momentum ΔH may be expressed as

$$\Delta H = \frac{E}{c}$$
$$= \frac{IA\Delta t}{c} \tag{6.2}$$

where A is the surface area and Δt is an interval of time. The magnitude of the force acting on the surface is, according to Newton's second law

$$F = \frac{\Delta H}{\Delta t}$$
$$= \frac{IA}{c}$$
$$= PA \tag{6.3}$$

where $P = I/c$ is the radiation pressure impinging on an energy absorbing surface. For a totally absorbing surface (e.g., a black body), $P = 4.5 \times 10^{-6} \ N/m^2$. For a perfectly reflecting surface (e.g., a mirror), $P = 9 \times 10^{-6} \ N/m^2$, which is greater by a factor of two due to the rebounding effect of the photons from the reflecting surface.

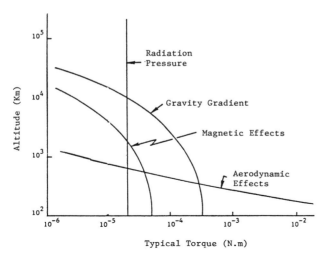

Figure 6.1 Typical torques on a small satellite as a function of altitude.

Spacecraft surfaces may reflect the impinging energy specularly or diffusely, as is illustrated in Figure 6.2.

In the case of specular reflection (Figure 6.2a), the angle of incidence with respect to the normal to the surface is equal to the angle of reflection. In diffuse reflection (Figure 6.2b), the radiation leaves the surface so that the intensity in any direction is proportional to the cosine of the angle between the surface normal and the direction of reflected radiation. A fraction of the incoming radiation may be absorbed and reflected diffusely, as is illustrated in Figure 6.2c.

6.1.1 Radiation Force Under Specular Reflection

Consider an element of area dA oriented at an angle θ relative to the incoming radiation, as shown in Figure 6.3. The solar radiation differential force components in terms of coefficient of reflection β are

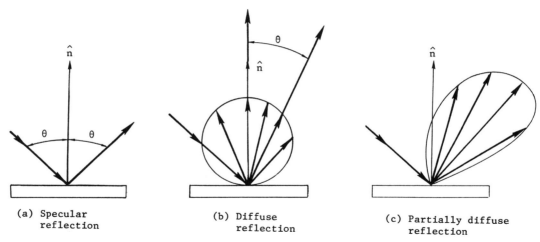

(a) Specular reflection   (b) Diffuse reflection   (c) Partially diffuse reflection

Figure 6.2 Types of reflection.

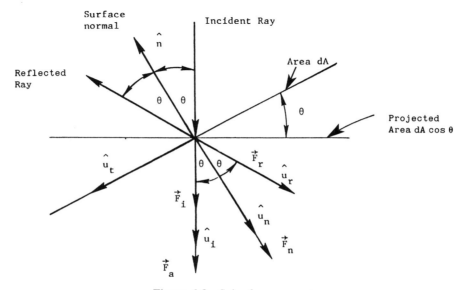

Figure 6.3 Solar force geometry.

$$d\vec{F}_i = \beta P \, dA \cos \theta \, \hat{u}_i \qquad (6.4)$$
$$= \text{force due to incident ray}$$
which is specularly reflected

$$d\vec{F}_a = (1 - \beta) P \, dA \cos \theta \, \hat{u}_i \qquad (6.5)$$
$$= \text{force due to absorbed incident ray}$$

$$d\vec{F}_r = \beta P \, dA \cos \theta \, \hat{u}_r \qquad (6.6)$$
$$= \text{force due to specularly reflected ray .}$$

Here the coefficient of reflection β is the reflected fraction of the solar radiation constant $I (0 \leq \beta \leq 1)$. Total force on an element of area dA for specular reflection is

$$\begin{aligned} d\vec{F}_{sp} &= d\vec{F}_a + d\vec{F}_i + d\vec{F}_r \\ &= d\vec{F}_a + d\vec{F}_n \end{aligned} \qquad (6.7)$$

where

$$\begin{aligned} d\vec{F}_n &= (d\vec{F}_r + d\vec{F}_i)\cos \theta \\ &= 2\beta P \, dA \cos^2 \theta \, \hat{u}_n . \end{aligned} \qquad (6.8)$$

6.1.2 Radiation Force Under Diffuse Reflection

If only a subfraction $s\beta$ of the reflected ray is reflected specularly, then $(1 - s)\beta$ is reflected diffusely. Here $0 \leq s \leq 1$. For pure specular reflection $s = 1$. For completely diffuse reflection $s = 0$. The force due only to the specularly reflected ray is then

$$d\vec{F}_s = s \, d\vec{F}_n . \qquad (6.9)$$

The force due to the stopping of the incoming ray is

$$d\vec{F}_{di} = (1 - s)\beta P \, dA \cos \theta \hat{u}_i . \qquad (6.10)$$

The force due to diffuse reflection of the $(1 - s)\beta$ fraction is

$$d\vec{F}_{dr} = (1 - s)\beta \frac{2P}{3} dA \cos \theta \hat{u}_n . \qquad (6.11)$$

The total force is

$$d\vec{f} = d\vec{F}_s + d\vec{F}_{di} + d\vec{F}_{dr} + d\vec{F}_a \qquad (6.12)$$

or

$$d\vec{f} = \left\{ \left[\frac{2\beta}{3}(1 - s)\cos \theta + (1 + s\beta)\cos^2 \theta \right] \hat{u}_n \right.$$
$$\left. + [(1 - s\beta)\cos \theta \sin \theta]\hat{u}_t \right\} P \, dA \qquad (6.13)$$

where $\hat{u}_t$ is a tangential unit vector along the surface direction (that is, $\hat{u}_t$ is orthogonal to the surface normal).

6.1.3 Radiation Force Limiting Cases

Elemental solar force expressions for the totally absorbing, specularly reflecting, and diffusely reflecting surfaces shown are:

a. Absorbing surface only

$$d\vec{f}_a = [\cos \theta \, \hat{u}_n + \sin \theta \, \hat{u}_t]\cos \theta P \, dA \qquad (6.14)$$

For this case, $\beta = 0$.

b. Specularly reflecting surface only

$$d\vec{f}_{rs} = [(1 + \beta)\cos \theta \, \hat{u}_n$$
$$+ (1 - \beta)\sin \theta \, \hat{u}_t]\cos \theta P \, dA \qquad (6.15)$$

where β = coefficient of specular reflection. For this case, $s = 1$.

c. Diffusely reflecting surface only

$$d\vec{f}_{rd} = \left[\left(\frac{2\beta_d}{3} + \cos \theta \right) \hat{u}_n \right.$$
$$\left. + \sin \theta \, \hat{u}_t \right] \cos \theta P \, dA \qquad (6.16)$$

For this case, β_d = coefficient of diffuse reflection and $s = 0$.

6.1.4 Radiation Torque

The torque due to solar radiation pressure arises due to the impingement of solar photons on the various distinct surfaces of the satellite. The incident photons thus produce a net force on each distinctive segment of the satellite. If the sum of all these forces does not pass through the center of mass of the satellite, which is ordinarily the case, a torque will be produced. For design purposes, the resulting solar radiation forces are assumed to pass through a "center of pressure" (CP). If the center of mass (CM) of the satellite can be located, the separation between the CP and the CM provides a useful measure of the effect of solar radiation torques since the CP-CM distance represents the moment arm through which the resultant force acts.

The main factors which affect the magnitude of the solar disturbance torque are:

a. Reflection characteristics of satellite surfaces.
b. Shading effects. Various parts of the satellite will not be continuously exposed during orbital motion since portions of the satellite structure will shield others from the sunlight. Satellites with large protruding parts will experience significant effects from shading.
c. Satellite configuration. The configuration influences the degree of shading and the CP-CM offset.
d. Orbital position and orbital inclination with respect to the Sun. The time history of the satellite in orbit affects the shading and the total amount of radiation reflected.

In general, the solar radiation torque on a spacecraft is of the form [2]

$$\vec{T}^{(s)} = \int_{\text{e.s.}} \vec{r} \times d\vec{f} \qquad (6.17)$$

where $\vec{r}$ is the vector directed from the spacecraft mass center to the element of area dA_i for the i-th surface, and e.s. denotes exposed surface. The latter can be defined in terms of the scalar product

$$\hat{n} \cdot \hat{\sigma} > 0 \qquad (6.18)$$

where $\hat{n}$ and $\hat{\sigma}$ are the surface normal and Sun direction unit vectors, respectively.

6.1.5 Example—Radiation Torque on a Geosynchronous Satellite

Consider the spacecraft of Figure 5.6. The geometry of the spacecraft in geosynchronous orbit is illustrated in Figure 6.4. Typical torques about the e_1, e_2, and e_3 body axes can be expressed as

$$T_1^{(s)} = a \cos S' + b \sin S' \cos \mu$$
$$T_2^{(s)} = c \sin S' \cos \mu$$
$$T_3^{(s)} = d \cos S' - e \sin S' \cos \mu \qquad (6.19)$$

where S' is the angle between the Sun line and the Earth's polar axis. Typical values for the five coefficients are on the order of 10^{-6} Newton-meter.

6.1.6 Momentum Change

For a given season (fixed value of S'), the fundamental Fourier components of solar torque about the body axes are, in general, of the form

$$T_1^{(s)} = A + B \cos \mu + G \sin \mu$$
$$T_2^{(s)} = E + F \cos \mu + I \sin \mu$$
$$T_3^{(s)} = C + D \sin \mu + H \cos \mu \qquad (6.20)$$

where $\mu = \omega_o t$, the in-orbit angle and where A through I are constants.

The angular momentum imparted to the spacecraft by solar radiation torques can be obtained by integrating the body torque components in an inertial (nonrotating) reference frame. The latter are given by the transformation

$$\begin{pmatrix} T_{1i} \\ T_{2i} \\ T_{3i} \end{pmatrix} = \begin{pmatrix} \cos \mu & 0 & -\sin \mu \\ 0 & 1 & 0 \\ \sin \mu & 0 & \cos \mu \end{pmatrix} \begin{pmatrix} T_1^{(s)} \\ T_2^{(s)} \\ T_3^{(s)} \end{pmatrix} \qquad (6.21)$$

Performing the matrix multiplication yields

$$T_{1i} = T_1^{(s)} \cos \mu - T_3^{(s)} \sin \mu$$
$$T_{2i} = T_2^{(s)}$$
$$T_{3i} = T_1^{(s)} \sin \mu + T_3^{(s)} \cos \mu \qquad (6.22)$$

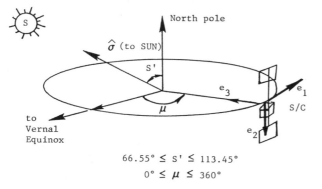

$$66.55° \le S' \le 113.45°$$
$$0° \le \mu \le 360°$$

Figure 6.4 Geosynchronous orbit geometry.

The secular (nonperiodic) components of the torques determine the amount of propellant required to remove the angular momentum imparted to the spacecraft. The form of the secular components can be obtained by inspection of equation 6.22 as

$$T_{1is} = B(\cos \mu)^2 - D(\sin \mu)^2$$
$$T_{2is} = E$$
$$T_{3is} = G(\sin \mu)^2 + H(\cos \mu)^2 . \qquad (6.23)$$

The mean values for T_{1is} and T_{3is} averaged over one orbital revolution are

$$\overline{T}_{1is} = \frac{1}{2}(B - D)$$

$$\overline{T}_{3is} = \frac{1}{2}(G + H) . \qquad (6.24)$$

The amplitude of the averaged secular solar radiation torque component per revolution in the $e_1 e_3$ plane is

$$\overline{T}_{13s} = (\overline{T}_{1is}^2 + \overline{T}_{3is}^2)^{1/2}$$
$$= \frac{1}{2}[(B - D)^2 + (G + H)^2]^{1/2} . \qquad (6.25)$$

The averaged angular momentum imparted to the spacecraft by the secular component of the solar radiation torque in the $e_1 e_3$ plane is then given by the equation

$$H_{13} = \int_o^{2\pi/\omega_0} \overline{T}_{13s} \, dt$$
$$= \frac{\pi}{\omega_o}[(B - D)^2 + (G + H)^2]^{1/2} . \qquad (6.26)$$

The angular momentum imparted to the e_2 axis is

$$H_2 = \frac{2\pi E}{\omega_o} \qquad (6.27)$$

which tends to speed up the RW aligned with its spin axis along the e_2 axis of the satellite.

A fraction of the angular momentum H_{13}, not entirely removed by thrusters, will also tend to increase the spin speed of the RW. This requires periodic dumping of the wheel angular momentum when a maximum speed is obtained.

6.2 Gravitational and Inertial Gradients

6.2.1 Gravitational Field

Newton's law of gravitation states that a spherical mass M creates a gravitational field in space around it which interacts with other masses. For a mass m at a distance R from the mass M, the gravitational potential energy (function) is given by

$$V = -\frac{GMm}{R} \qquad (6.28)$$

where the universal gravitational constant $G = 6.6726 \times 10^{-11}$ m^3/kg sec^2. The gradient of the potential is the gravitational force field. Thus, the gravitational acceleration on the mass m can be defined as

$$\vec{a}_g = \frac{\nabla_\alpha V}{m} \qquad (6.29)$$

where

$$\nabla_\alpha = \hat{e}_1 \frac{\partial}{\partial e_1} + \hat{e}_2 \frac{\partial}{\partial e_2} + \hat{e}_3 \frac{\partial}{\partial e_3}$$

in terms of the coordinates given in Figure 6.5.

The components of the gradient of the gravitational field in a cartesian coordinate frame are given by [3, 4] as:

$$\Gamma_{\alpha\beta} = \nabla_\alpha \nabla_\beta \frac{V}{m} \qquad (\alpha, \beta = e_1, e_2, e_3)$$

$$\Gamma_{\alpha\beta} = \frac{1}{m} \begin{pmatrix} \dfrac{\partial^2 V}{\partial e_1^2} & \dfrac{\partial^2 V}{\partial e_1 \partial e_2} & \dfrac{\partial^2 V}{\partial e_1 \partial e_3} \\ \dfrac{\partial^2 V}{\partial e_1 \partial e_2} & \dfrac{\partial^2 V}{\partial e_2^2} & \dfrac{\partial^2 V}{\partial e_2 \partial e_3} \\ \dfrac{\partial^2 V}{\partial e_1 \partial e_3} & \dfrac{\partial^2 V}{\partial e_2 \partial e_3} & \dfrac{\partial^2 V}{\partial e_3^2} \end{pmatrix}$$

$$(6.30)$$

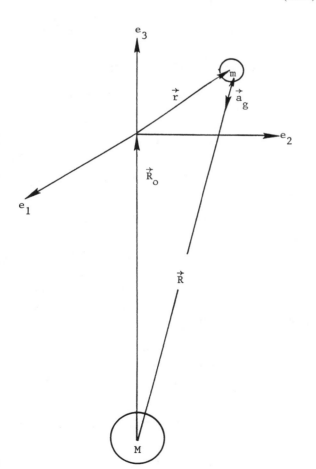

Figure 6.5   Inertial reference frame.

Here $\Gamma_{\alpha\beta}$ is the gravitational acceleration gradient which causes an acceleration in the α direction of an object displaced in the β direction. Thus, the acceleration in the e_1 direction is given by

$$a_1 = \Gamma_{11} e_1 + \Gamma_{12} e_2 + \Gamma_{13} e_3 \qquad (6.31)$$

Conversely

$$\Gamma_{11} = \frac{\partial a_1}{\partial e_1}, \quad \Gamma_{12} = \frac{\partial a_1}{\partial e_2}, \quad \Gamma_{13} = \frac{\partial a_1}{\partial e_3}. \qquad (6.32)$$

In an inertial reference frame, as shown in Figure 6.5, the gravitational acceleration can be expressed as follows:

$$\begin{aligned} \vec{a}_g &= -\frac{GM\vec{R}}{R^3} \\ &= -\frac{GM(\vec{R}_o + \vec{r})}{R^3} \\ &= -\frac{GM(\vec{R}_o + \vec{r})}{[(\vec{R}_o + \vec{r}) \cdot (\vec{R}_o + \vec{r})]^{3/2}} \\ &= -\frac{GM(\vec{R}_o + \vec{r})}{R_o^3} \\ &\quad \times \left[1 + \left(\frac{r}{R_o}\right)^2 + \frac{2\vec{R}_o \cdot \vec{r}}{R_o^2} \right]^{-3/2} \\ &= -\frac{GM}{R_o^3} \left[\vec{R}_o + \vec{r} - 3\frac{\vec{R}_o \cdot \vec{r}}{R_o^2} \vec{R}_o \right] \\ &\quad + \text{higher order terms} . \end{aligned} \qquad (6.33)$$

Therefore, if

$$\vec{R}_o = R_o \hat{e}_3 \qquad (6.34)$$

and

$$\vec{r} = r_1 \hat{e}_1 + r_2 \hat{e}_2 + r_3 \hat{e}_3 \qquad (6.35)$$

then

$$\vec{a}_g \approx -\frac{GM}{R_o^3} [r_1 \hat{e}_1 + r_2 \hat{e}_2 + (R_o - 2r_3)\hat{e}_3] \qquad (6.36)$$

yielding

$$\Gamma_{11} = \frac{\partial a_g}{\partial r_1} = -\frac{GM}{R_o^3} \qquad (6.37)$$

$$\Gamma_{22} = \frac{\partial a_g}{\partial r_2} = -\frac{GM}{R_o^3} \qquad (6.38)$$

$$\Gamma_{33} = \frac{\partial a_g}{\partial r_3} = \frac{2GM}{R_o^3} . \qquad (6.39)$$

Consequently, the gravitational gradient matrix is of the form

$$\Gamma = \frac{GM}{R_o^3} \begin{pmatrix} -1 & 0 & 0 \\ 0 & -1 & 0 \\ 0 & 0 & 2 \end{pmatrix}. \qquad (6.40)$$

6.2.2 Inertial Gradient Field

If the reference frame is not an inertial or Newtonian frame but has an angular velocity $\vec{\omega}$, then an inertial (rotational) acceleration field is developed. The gradients from this field can be added to the gravitational gradients to obtain the total gradient.

The general expression for the acceleration of a point mass m in an inertial reference frame, as derived in Chapter 1, is

$$\vec{a} = \vec{a}_o + \ddot{\vec{r}}' + \vec{\omega} \times (\vec{\omega} \times \vec{r})$$
$$+ \dot{\vec{\omega}} \times \vec{r} + 2\vec{\omega} \times \dot{\vec{r}}' \quad (6.41)$$

where

$\vec{a}_o$ = acceleration of the origin of the rotating reference frame

$\ddot{\vec{r}}'$ = acceleration (relative) in the rotating reference frame

$\vec{\omega}$ = angular velocity of the rotating reference frame

$\dot{\vec{r}}'$ = velocity (relative) in the rotating reference frame

$\vec{r}$ = position vector in the rotating frame.

In the rotating frame, the acceleration is

$$\ddot{\vec{r}}' = \vec{a} - [\vec{a}_o + \vec{\omega} \times (\vec{\omega} \times \vec{r})$$
$$+ \dot{\vec{\omega}} \times \vec{r} + 2\vec{\omega} \times \dot{\vec{r}}'] . \quad (6.42)$$

Only terms involving $\vec{r}$ have gradients which can be computed and added to the gravitational gradients in equation 6.40. Thus, let

$$\vec{A} = -[\vec{\omega} \times (\vec{\omega} \times \vec{r}) + \dot{\vec{\omega}} \times \vec{r}]$$
$$= \omega^2 \vec{r} - \vec{\omega}(\vec{\omega} \cdot \vec{r}) + \vec{r} \times \dot{\vec{\omega}}$$
$$= A_1 \hat{e}_1 + A_2 \hat{e}_2 + A_3 \hat{e}_3 \quad (6.43)$$

where

$$A_1 = r_2\dot{\omega}_3 - r_3\dot{\omega}_2 + (\omega_2^2 + \omega_3^2)r_1 - \omega_1\omega_2 r_2 - \omega_1\omega_3 r_3 \quad (6.44)$$
$$A_2 = r_3\dot{\omega}_1 - r_1\dot{\omega}_3 + (\omega_3^2 + \omega_1^2)r_2 - \omega_2\omega_3 r_3 - \omega_2\omega_1 r_1 \quad (6.45)$$
$$A_3 = r_1\dot{\omega}_2 - r_2\dot{\omega}_1 + (\omega_1^2 + \omega_2^2)r_3 - \omega_3\omega_1 r_1 - \omega_3\omega_2 r_2 \quad (6.46)$$

and, as before,

$$\vec{\omega} = \omega_1 \hat{e}_1 + \omega_2 \hat{e}_2 + \omega_3 \hat{e}_3 . \quad (6.47)$$

Various partial derivatives are then of the form

$$A_{11} = \frac{\partial A_1}{\partial r_1} , \quad A_{12} = \frac{\partial A_1}{\partial r_2} , \ldots .$$

They yield the composite matrix of the inertial and gravitational gradient terms as follows [3]:

$$G_{\alpha\beta} = \begin{pmatrix} (\Gamma_{11} + \omega_2^2 + \omega_3^2) & (\dot{\omega}_3 - \omega_1\omega_2) & -(\dot{\omega}_2 + \omega_1\omega_3) \\ -(\dot{\omega}_3 + \omega_1\omega_2) & (\Gamma_{22} + \omega_1^2 + \omega_3^2) & (\dot{\omega}_1 - \omega_2\omega_3) \\ (\dot{\omega}_2 - \omega_1\omega_3) & -(\dot{\omega}_1 + \omega_2\omega_3) & (\Gamma_{33} + \omega_1^2 + \omega_2^2) \end{pmatrix}$$
$$(6.48)$$

6.2.3 Circular Orbit Case

For a circular orbit $\omega_o^2 = GM/R_o^3$, where ω_o is the angular velocity of the orbit (or the reference frame). If $\hat{e}_1$ is directed along the orbit normal, then

$$\omega_1 = \omega_o \quad \text{and} \quad \omega_2 = \omega_3 = 0 .$$

The gravitational and inertial gradients along the e_1, e_2, and e_3 axes are illustrated in Figures 6.6(a) and 6.6(b), respectively.

The combined gravitational and centrifugal (rotational) gradients in an orbit oriented (rotating) reference frame are shown in Figure 6.7.

In matrix notation, the combined gradient field is of the form

$$G_{\alpha\beta} = \omega_o^2 \begin{pmatrix} -1 & 0 & 0 \\ 0 & 0 & 0 \\ 0 & 0 & 3 \end{pmatrix} . \quad (6.49)$$

The significance of the above result is that masses along the e_1 axis are subject to compressive loads (accelerations). Masses along the e_2 axis experience no net accelerations but along the e_3 axis they experience a net tension (along the local vertical direction). These characteristics imply structural stability of a flexible body (e.g., a tether or a chain) deployed along the local vertical in a circular orbit.

6.3 Gravity Gradient Torque

A gravitational gradient torque acting on a distributed mass body in orbit is given by the expression

$$\vec{T}^{(g)} = \int_m \vec{r} \times \vec{a}_g \, dm$$
$$\approx -\frac{GM}{R_o^3} \int_m \vec{r}$$
$$\times \left[\vec{R}_o + \vec{r} - 3\frac{\vec{R}_o \cdot \vec{r}}{R_o^2} \vec{R}_o \right] dm$$
$$\approx \frac{3GM}{R_o^3} \int_m \vec{r} \times \frac{\vec{R}_o \cdot \vec{r}}{R_o^2} \vec{R}_o \, dm . \quad (6.50)$$

Equation 6.50 can be integrated as follows. Let $\mu = GM$, and $\vec{R}_o = R_o\hat{E}_1$, where $\hat{E}_1$ is the unit vector along the outward radius. Then

$$\vec{T}^{(g)} \approx \frac{3\mu}{R_o^3} \int_m \frac{(\vec{R}_o \cdot \vec{r})\vec{r} \times \vec{R}_o}{R_o^2} \, dm$$
$$= \frac{3\mu}{R_o^3} \hat{E}_1 \cdot \int_m \vec{r}\vec{r} \, dm \times \hat{E}_1$$
$$= \frac{3\mu}{R_o^3} \hat{E}_1 \times \int_m (\bar{\bar{E}} r^2 - \vec{r}\vec{r}) dm \cdot \hat{E}_1$$
$$= \frac{3\mu}{R_o^3} \hat{E}_1 \times \bar{\bar{I}} \cdot \hat{E}_1 \quad (6.51)$$

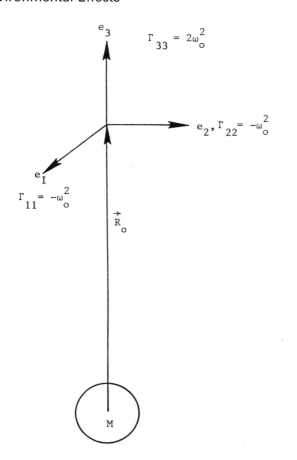

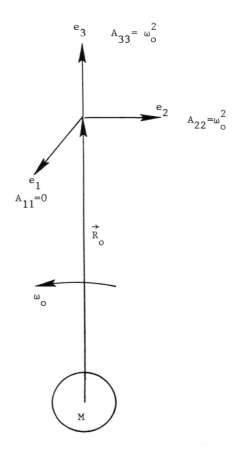

(a) Gravitational gradients   (b) Centrifugal gradients

Figure 6.6 Gravitational and centrifugal gradients between masses M and m.

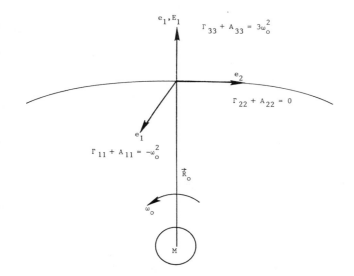

Figure 6.7 The combined gravitational and centrifugal gradient field.

where $\bar{\bar{E}} = \hat{E}_1\hat{E}_1 + \hat{E}_2\hat{E}_2 + \hat{E}_3\hat{E}_3$ is a unit dyadic, and where, as derived in Chapter, 1, the inertia dyadic about the body's center of mass (origin of reference frame) is

$$\bar{\bar{I}} = \int_m (\bar{\bar{E}}r^2 - \vec{r}\vec{r})dm .$$

With respect to the satellite body axes $e_\alpha (\alpha = 1, 2, 3)$, the gravity gradient torque becomes

$$\vec{T}^{(g)} = K\hat{E}_1 \times \bar{\bar{I}} \cdot \hat{E}_1$$
$$= Ka_{\alpha 1}\hat{e}_\alpha \times I_{\alpha\beta}\hat{e}_\alpha\hat{e}_\beta \cdot a_{\alpha 1}\hat{e}_\alpha \qquad (6.52)$$

where

$$\hat{E}_1 = a_{\alpha 1}\hat{e}_\alpha = a_{11}\hat{e}_1 + a_{21}\hat{e}_2 + a_{31}\hat{e}_3$$
$$\bar{\bar{I}} = I_{\alpha\beta}\hat{e}_\alpha\hat{e}_\beta \qquad (\alpha, \beta = 1, 2, 3)$$
$$K = 3\mu/R_o^3$$

and $a_{\alpha 1}$ are the direction cosines between the $\hat{E}_1$ and $\hat{e}_\alpha$ unit vectors.

Equation 6.52 can be written in scalar form to yield the body components of torque as follows:

$$T_\lambda^{(g)} = \vec{T}^{(g)} \cdot \hat{e}_\lambda \qquad (\lambda = 1, 2, 3)$$
$$= K a_{\alpha 1} \, a_{\beta 1} \, I_{\gamma \beta} \, \varepsilon_{\alpha \gamma \lambda} \qquad (6.53)$$

where the three dimensional permutation tensor, as defined in Chapter 1, is

$$\varepsilon_{\alpha \gamma \lambda} = (\hat{e}_\alpha \times \hat{e}_\gamma) \cdot \hat{e}_\lambda \qquad (\alpha, \beta, \gamma, \lambda = 1, 2, 3)$$

For the principal body axes, $I_{\gamma \beta} = 0$ for $\gamma \neq \beta$, and the torque components become

$$T_1^{(g)} = K(I_{33} - I_{22}) a_{21} a_{31}$$
$$T_2^{(g)} = K(I_{11} - I_{33}) a_{11} a_{31}$$
$$T_3^{(g)} = K(I_{22} - I_{11}) a_{11} a_{21} \qquad (6.54)$$

where the $\alpha_{\alpha\beta}$ terms are the elements (i.e., direction cosines) of the transformation matrix. That is $\hat{e}_\alpha = a_{\alpha\beta} \hat{E}_\beta$. For example, the sequential rotations in Figure 1.7 yield the rotation matrix R_{123} in equation 1.10. Therefore, the gravitational torques about the principal body axes e_1, e_2, and e_3 are of the form

$$T_1^{(g)} = -K(I_{33} - I_{22}) c\theta_2 s\theta_3 s\theta_2$$
$$= 0 \quad \text{for} \quad \theta_2 = \theta_3 = 0$$
$$T_2^{(g)} = K(I_{11} - I_{33}) c\theta_3 c\theta_2 s\theta_2$$
$$T_3^{(g)} = -K(I_{22} - I_{11}) c\theta_3 (c\theta_2)^2 s\theta_3 . \qquad (6.55)$$

6.4 Geomagnetic Field

The geomagnetic field can be represented to a first order approximation as a magnetic dipole with the axis inclined to the Earth's spin axis by approximately 11.5 degrees. Observations have shown that the sources of the geomagnetic field are in the core and at the surface of the Earth as well as in the upper atmosphere. The primary source is presumed to be a system of electric currents in the Earth's molten interior. The changes in these currents account for the migration of the geomagnetic poles on the Earth's surface.

Ferromagnetic minerals and metals in various locations on the Earth's surface provide a second source of geomagnetism. This source accounts for certain "magnetic anomalies" which can cause deviations from the dipole field.

A third source of the magnetic field is associated with the motion of charged particles in the upper atmosphere and near Earth space. The movement of these positive and negative ions creates magnetic fields of various intensities. When these charged particles enter the Earth's upper atmosphere they are responsible for the so-called northern and southern lights.

The geomagnetic field vector $\vec{B}$ can be expressed in the form

$$\vec{B} = \vec{B}_o + \vec{B}_r + \vec{B}_a + \vec{B}_c + \delta\vec{B} \qquad (6.56)$$

where

$\vec{B}_o$ = homogeneous field (90% of $\vec{B}$)

$\vec{B}_r$ = continental (remanent) field

$\vec{B}_a$ = anomalous field (up to 5 km above Earth's surface)

$\vec{B}_e$ = charged particles field

δB = variational component due to solar wind variations and cosmic rays

Therefore, $\vec{B}$ is a function of altitude, longitude, and latitude and can be expressed in units of gauss, gamma (γ), or tesla (webers per square meter). A one gauss magnetic field intensity acting on a unit magnetic pole produces a force of one dyne.

The various units of magnetic field intensity or magnetic induction are related as follows:

$$1 \text{ tesla} = 1 \text{ weber}/m^2$$
$$= 10^4 \text{ gauss}$$
$$= 10^9 \text{ gamma} (\gamma)$$

The tesla has SI dimensions of kilogram $\cdot$ ampere^{-1} $\cdot$ second^{-2}.

6.4.1 The Near Field Effects

The $\vec{B}_o$ component of the geomagnetic field may be represented as shown in Figure 6.8. The field $\vec{B}_o$ at a distance $\vec{r}$ from the center of a bar magnet is given by

$$\vec{B}_o = -\nabla\phi \qquad (6.57)$$

where $\phi = \mu_E \cos\theta/r^2$ is the dipole potential and $\mu_E = 8.1 \times 10^{25}$ gauss $\cdot$ cm^3 is the magnitude of the Earth's magnetic moment vector along the magnet axial direction. ∇ is the gradient operator in polar coordinates and is expressed as

$$\nabla = \hat{e}_r \frac{\partial}{\partial r} + \hat{e}_\theta \frac{1}{r} \frac{\partial}{\partial\theta} .$$

Here θ is the angle measured from the bar magnet axis to the radius vector $\vec{r}$, and $\hat{e}_r$, $\hat{e}_\theta$ are the unit vectors along the radius and normal to it, respectively. Therefore,

$$\vec{B}_o = \frac{\mu_E}{r^3} (2 \cos\theta \hat{e}_r + \sin\theta \hat{e}_\theta)$$
$$= \frac{\mu_E}{r^3} (-2 \sin\theta_m \hat{e}_r + \cos\theta_m \hat{e}_\theta) \qquad (6.58)$$

where θ_m is the magnetic latitude measured from the geomagnetic equator. The magnitude of the field is therefore

$$|\vec{B}_o| = \frac{\mu_E}{r^3} (1 + 3 \sin^2\theta_m)^{1/2} \qquad (6.59)$$

At the Earth's surface along the geomagnetic equator, $\theta_m = 0$, and $B_o = 0.311$ gauss. At the magnetic pole, $B_o = 0.622$ gauss or twice the equatorial value.

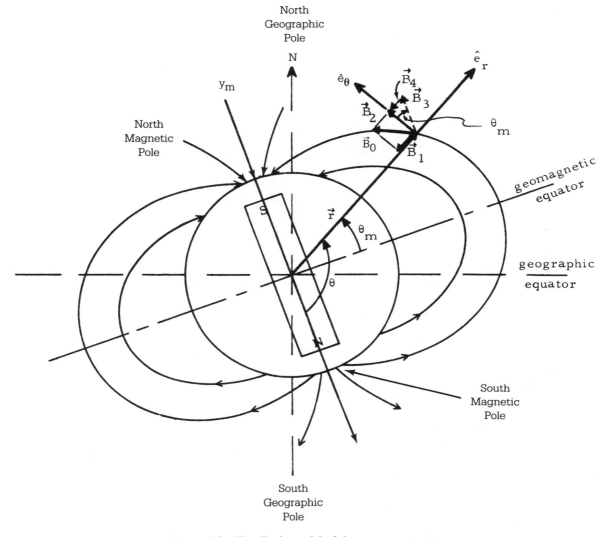

Figure 6.8 The dipole model of the geomagnetic field.

The field $\vec{B}_o$ can also be expressed in terms of the radial ($\hat{e}_r$) and axial ($\hat{y}_m$) components. By inspection of Figure 6.8

$$\begin{aligned} \vec{B}_o &= \vec{B}_1 + \vec{B}_2 \\ &= \vec{B}_1 + \vec{B}_3 + \vec{B}_4 . \end{aligned} \qquad (6.60)$$

The individual components are

$$\vec{B}_1 = -2 \frac{\mu_E}{r^3} \sin \theta_m \hat{e}_r$$

$$\vec{B}_2 = \frac{\mu_E}{r^3} \cos \theta_m \hat{e}_\theta$$

$$\vec{B}_3 = \frac{B_2}{\cos \theta_m} \hat{y}_m = \frac{\mu_E}{r^3} \hat{y}_m$$

$$\vec{B}_4 = -B_3 \sin \theta_m \hat{e}_r = -\frac{\mu_E}{r^3} \sin \theta_m \hat{e}_r .$$

Therefore,

$$\vec{B}_o = -\frac{\mu_E}{r^3} (3 \sin \theta_m \hat{e}_r - \hat{y}_m) . \qquad (6.61)$$

This is equivalent to

$$\vec{B}_o = \frac{\vec{\mu}_E}{r^3} - 3 \frac{(\vec{\mu}_E \cdot \vec{r})\vec{r}}{r^5} \qquad (6.62)$$

where $\vec{\mu}_E = \mu_E \hat{y}_m$ is the Earth's magnetic moment vector.

To find the components of $\vec{B}_o$ in the satellite body axes e_α, we first resolve $\vec{B}_o$ in the orbit axes $E_\beta(\alpha, \beta = 1, 2, 3)$ as shown in Figure 6.9. Since $\vec{B}_1 + \vec{B}_4$ are along the radial $\hat{e}_r$ direction which corresponds to E_1; therefore, only the B_3 component needs to be resolved in the E_β frame. This can be done by rotation through the four angles α_1 through α_4

of Figure 6.9 and adding the resulting components $\vec{B}_1$ + $\vec{B}_4$. For example,

$$\begin{pmatrix} B_{E_1} \\ B_{E_2} \\ B_{E_3} \end{pmatrix} = \begin{pmatrix} B_1 + B_4 \\ 0 \\ 0 \end{pmatrix}$$

$$+ M_2(\alpha_4) M_1(\alpha_3) M_2(-\alpha_2) M_1(-\alpha_1) \begin{pmatrix} 0 \\ B_3 \\ 0 \end{pmatrix} \quad (6.63)$$

where B_{E_1}, B_{E_2}, and B_{E_3} are the magnetic field components along the E_1, E_2, and E_3 orbit reference axes, respectively, and where the rotation matrices are of the form

$$M_1(\mu) = \begin{pmatrix} 1 & 0 & 0 \\ 0 & c\mu & s\mu \\ 0 & -s\mu & c\mu \end{pmatrix}$$

$$M_2(\mu) = \begin{pmatrix} c\mu & 0 & -s\mu \\ 0 & 1 & 0 \\ s\mu & 0 & c\mu \end{pmatrix} .$$

The component sum $B_1 + B_4$ in equation 6.63 is

$$B_1 + B_4 = -\frac{3\mu_E}{r^3} \sin \theta_m . \quad (6.64)$$

The spacecraft body components of the magnetic field vector are given by

$$\begin{pmatrix} B_{e_1} \\ B_{e_2} \\ B_{e_3} \end{pmatrix} = R \begin{pmatrix} B_{E_1} \\ B_{E_2} \\ B_{E_3} \end{pmatrix} \quad (6.65)$$

where the rotation matrix R relates the spacecraft body axes $e_\alpha(\alpha = 1, 2, 3)$ relative to the orbit reference axes $E_\beta(\beta = 1, 2, 3)$.

A plot of the Earth's equatorial magnetic field intensity B_{eq} in gamma (γ) units as a function of altitude is shown in Figure 6.10. The effect of latitude is given in Figure 6.11. For any latitude $B_o = \alpha B_{eq}$.

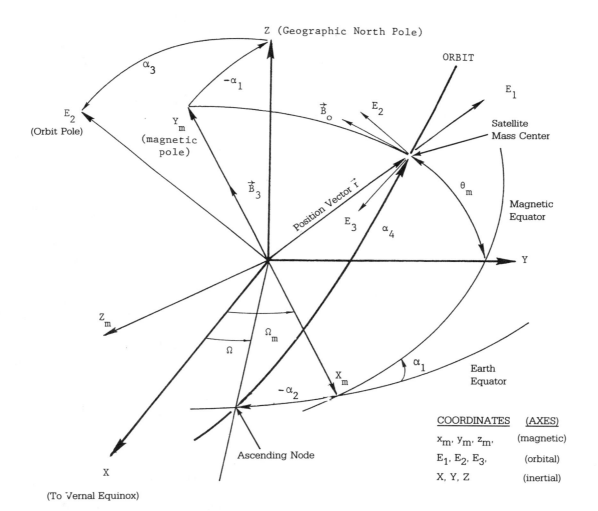

Figure 6.9 Geomagnetic and orbital coordinates.

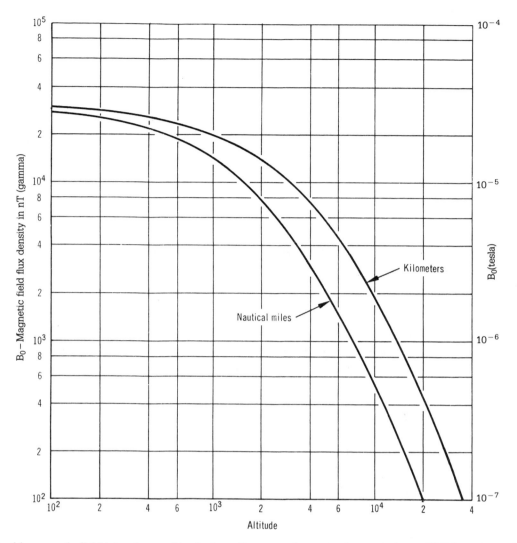

Figure 6.10   Earth's magnetic field intensity vs. altitude above the magnetic equator for $\mu_E = 1 \times 10^{17}$ Wb-m (8×10^{25} pole-cm) [5].

6.4.2   The Far Field Effects

The geomagnetic field at and beyond synchronous altitude (i.e., at 6.6 Earth radii and higher) is distinctly different from that of the near field. Analyses of satellite data have shown that the geomagnetic field can be divided into the following main regions, as illustrated in Figure 6.12: (1) the interplanetary region, (2) the interaction region, and (3) the magnetosphere.

6.4.3   The Interplanetary Region

In early magnetic investigations it was assumed that the interplanetary magnetic field was negligible and that immediately outside the magnetopause the solar plasma would have no magnetic field associated with it. However, recent studies have suggested, and satellite measurements have confirmed, the existence of a spiral magnetic field extending from the Sun to the vicinity of the Earth. This inter-

planetary region of the magnetic field begins at the collisionless shock wavefront, which develops in the supersonic flow of the solar wind as it interacts with the geomagnetic field, and extends indefinitely away from the Earth toward the Sun, as shown in Figure 6.12. Satellite measurements of the shock wave boundary, which separates the interplanetary and interaction regions, show the boundary's closest proximity to the Earth to be about 14 Earth radii from the Earth's center in the subsolar direction, i.e., in the sunlit direction of a line joining the centers of the Earth and the Sun. From this 14 Earth radii location at the subsolar point (noon meridian), the shock wave boundary tapers away from the Earth to about 22 Earth radii along the dusk and dawn meridians as viewed along the ecliptic plane [6].

Satellite magnetometer measurements have also shown the magnetic field in the interplanetary region to have an average intensity that is relatively constant at the 5 to 10

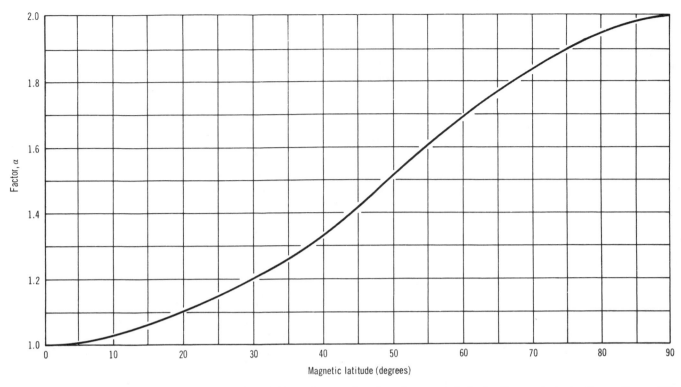

Figure 6.11   Multiplication factor α for magnetic field intensity as a function of magnetic latitude. For any latitude $B_o = \alpha B_{\text{equatorial}}$ [5].

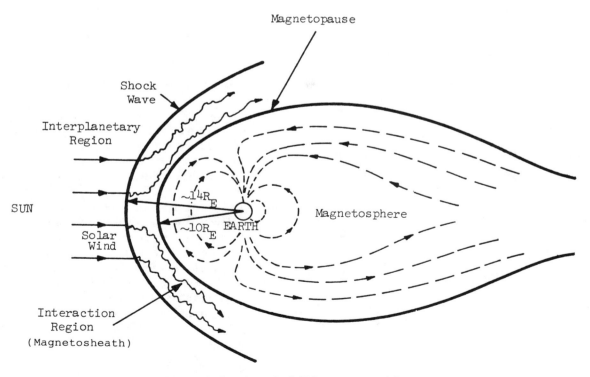

Figure 6.12   Magnetic field in extraterrestrial space.

gammas level and a direction that is somewhat random. Figure 6.13 presents data which is typical of magnetic measurements obtained by satellites. This particular recording clearly shows the shock wave boundary at about 14 Earth radii, where the magnetic intensity also drops to the interplanetary level of five gammas.

6.4.4 The Interaction Region

The magnetic region between the shock wave front and the magnetopause boundaries has been referred to as the magnetosheath or the interaction region. Here, the interaction of the solar wind with the interplanetary and geomagnetic fields causes the resultant magnetic field to change abruptly and become very turbulent or dynamic. Although numerous satellite measurements have shown the field in this region to range between 10 and 50 gammas and to exhibit large fluctuations in both strength and magnitude, postulated causes of the turbulence by eminent geophysicists still remain unproven.

6.4.5 The Magnetosphere

Satellite measurements indicate that the geomagnetic field is distorted and confined to a definite volume of space called the magnetosphere, which is shaped somewhat like an enormous teardrop with a tremendously long tail. The pressure of the solar wind compresses the geomagnetic field on the Earth's sunlit side (subsolar direction) and bends the polar field lines back toward an elongated tail on the Earth's dark side (antisolar direction). The plasma is most effective in compressing the magnetosphere where the direction of the plasma flow is perpendicular to the surface of the magnetosphere boundary, the magnetopause. In the simplest situation, where the plasma is assumed to be streaming

radially from the Sun in straight lines, the point of maximum compression is along the line that connects the centers of the Earth and the Sun. Away from this line, e.g., toward either polar cap, the plasma strikes the magnetopause at less than right angles and, hence, is less effective in compressing the geomagnetic field lines. Moreover, on the dark side of the Earth (in the antisolar direction), the solar plasma no longer compresses the field lines; instead, tangential stresses cause the geomagnetic field lines of the magnetosphere to stretch out into long tails which are either closed or open.

The teardrop shaped magnetosphere is bounded by the magnetopause, which has been repeatedly observed and recorded by a number of satellites. On the basis of such measurements, it is possible to state with some assurance that the magnetosphere is nearly symmetrical about the Earth-Sun line and the nearest nominal magnetopause location is approximately 10 Earth radii from the Earth's center on the Earth's sunlit side. From this noon meridian location, the magnetopause's nominal location increases to a position of approximately 14 Earth radii at the dawn and dusk meridians. Then, the magnetopause stretches back into an elongated tail which extends at least 30 Earth radii or more on the Earth's dark side (antisolar direction or midnight meridian).

6.4.6 Magnetic Torques

Magnetic torques acting on a spacecraft can result from the interaction of the spacecraft's residual magnetic field and the geomagnetic field. Thus, if $\vec{M}$ is the sum of all magnetic moments in the spacecraft the torque acting on the spacecraft is

$$\vec{T}^{(m)} = \vec{M} \times \vec{B} \qquad (6.66)$$

where $\vec{B}$ = geomagnetic field vector. In general, $\vec{M}$ can be caused by permanent and induced magnetism or by spacecraft-generated current loops. The units of M may be gauss-cm³, ampere-m², or pole-cm. For example, if M is in ampere-m² and B in tesla or webers/m², then $T^{(m)}$ is in Newton-meters; or if M is in pole-cm and B in gauss, then $T^{(m)}$ is in dyne-cm. In general,

$$1 \text{ ampere-m}^2 = 1000 \text{ pole-cm}$$
$$1 \text{ pole-cm} = 1 \frac{\text{dyne-cm}}{\text{gauss}}.$$

The torques caused by the spinning motion of the spacecraft are induced by eddy currents, which then interact with any magnetized permeable spacecraft materials. Eddy current torques are of the form

$$\vec{T}^{(m)}_{\text{eddy}} = k_e (\vec{\omega} \times \vec{B}) \times \vec{B} \qquad (6.67)$$

where k_e is a constant and $\vec{\omega}$ is the spacecraft spin vector.

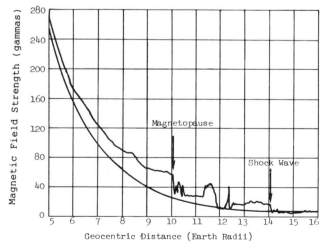

Figure 6.13 Typical satellite magnetic field strength measurements as a function of distance.

6.5 Aerodynamic Torques

The satellite will, in general, pass through an atmosphere of density ρ, with a velocity $\vec{v}$. The magnitude of the aerodynamic force $F^{(a)}$ is then given as

$$F^{(a)} = \frac{1}{2} \rho \vec{v} \cdot \vec{v} A C_d \qquad (6.68)$$

where A is the reference area of the satellite (such as the cross section along $\vec{v}$) and C_d is the total drag coefficient [7, 9]. Torque is given by

$$T^{(a)} = \frac{1}{2} \rho v^2 \ell S C_d \qquad (6.69)$$

where ℓ is the length of the perpendicular from the mass center to the force line of action.

Difficulties arise in determining ρ, ℓ, and C_d. At satellite altitudes, ρ is highly dependent on the time of day and the level of solar activity. For example, at 600 km the solar daytime maximum density and the nighttime minimum density may differ by a factor of 100. The following table gives typical values of the daytime maximum air density as a function of orbital height:

Height (km)	Density (kg/m^3)
200	4×10^{-10}
300	5×10^{-11}
400	1.5×10^{-11}
500	5×10^{-12}
600	2×10^{-12}
700	8×10^{-13}

For a satellite having a spherical shape, an average value of $C_d = 2.2$ can be taken which is computed assuming "free molecular flow"; i.e., the molecular mean (average) free path is assumed to be large compared to the size of the satellite and, therefore, interparticle collisions are ignored. For a cylinder C_d may be taken as 3.

The effects of magnetic, gravitational, and solar radiation torques on spacecraft performance are discussed in references 10 and 11. Also, additional information on these torques and on the Earth's atmosphere is given in references 12–15. A generalized vector model of the environmental torques is presented in reference 16.

6.6 References

1. Acord, J. D., and J. C. Nicklas. "Theoretical and Practical Aspects of Solar Pressure Attitude Control for Interplanetary Spacecraft," Jet Propulsion Laboratory, CA, TR No. 32–467, May 1964.

2. "Spacecraft Radiation Torques," NASA Space Vehicle Design Criteria, NASA SP-8027, October 1969.

3. Savet, P. H. "A New Type of Space Exploration by Gradient Technique," AIAA/GCFD Conference Preprint No. 68–851, Pasadena, CA, August 1968.

4. Roberson, R. E. "Gravity Gradient Determination of the Vertical," ARS Journal, November 1961.

5. "Spacecraft Magnetic Torques," NASA Space Vehicle Design Criteria, NASA SP-8018, March 1969.

6. Luke, R. K. C. "Description of the Geomagnetic Field," The Aerospace Corporation, TR-1001(2307)-5, January 1967.

7. "Spacecraft Aerodynamics Torques," NASA Space Vehicle Design Criteria, NASA SP-8058, January 1971.

8. DeBra, D. B. "The Effect of Aerodynamic Forces on Satellite Attitude," Advances Astron. Sci., Vol. 3, Paper 32, 1958.

9. Sentman, L. H., and S. E. Neice. "Drag Coefficients for Tumbling Satellites," Journal of Spacecraft and Rockets, Vol. 4, No. 9, September 1967.

10. Hara, M. "Effects of Magnetic and Gravitational Torques on Spinning Satellite Attitude," AIAA Journal, Vol. 11, No. 12, 1973, pp. 1737–1742.

11. Sincarsin, G. B., and P. C. Hughes. "Torques from Solar Radiation Pressure Gradient During Eclipse," Journal of Guidance, Control, and Dynamics, Vol. 6, No. 6, 1983, pp. 511–517.

12. "Models of Earth's Atmosphere (120–1000 km)," NASA SP-8021, May 1969.

13. "Spacecraft Gravitational Torques," NASA SP-8024, May 1969.

14. "Solar Electromagnetic Radiation," NASA SP8005, May 1971.

15. Tobiska, K., Culp, R. D., and C. A. Barth. "Prediction of Orbit Decay Through Next Solar Cycle Solar Mesosphere Explorer," AIAA Paper No. 86–2223, 1986.

16. Shivanaud, B. "Spacecraft Attitude Perturbation Torques due to Space Environmental Sources," AAS/AIAA Paper No. 85–329, 1985.

Chapter 7
Passive Gravity Gradient Stabilization

7.1 Introduction

The interaction between the satellite inertia ellipsoid and the gravitational field can be used to provide geocentric orientation of the satellite. Passive gravity gradient stabilization achieves this orientation for the spacecraft without the use of active control elements such as servo systems, RWs, or gas jets. Passive techniques can include moving parts which utilize the environment to damp oscillations (librations) about the reference axes.

Primary advantages of gravity gradient stabilization are long life, low power requirements, and Earth or planet pointing attitude capability. The disadvantages are low control torques, libration damper requirements, and imprecise pointing capabilities (e.g., 1 to 10° about all axes).

A rigid body in a circular orbit will tend to be aligned by the combined gravity gradient and centrifugal effects so that the axis of maximum moment of inertia is normal to the orbit plane (pitch axis). In addition, the axis of minimum moment of inertia is aligned along the local vertical and the intermediate principal axis lies along the velocity vector. A simple example of the gravity gradient restoring torque acting on the dumbbell shaped vehicle is illustrated in Figure 7.1. The torque on the dumbbell about the composite center of mass arises as a result of the inverse relationship between the gravity gradient and the distance to the center of the Earth. The reaction of the vehicle to the gravitational and also centrifugal forces imposed is such that all forces acting on the mass center are equalized. The velocity of the lower mass is smaller than that of the upper mass. In addition, the gravity force acting on the lower mass is greater than the gravity force on the more distant upper mass. The net result is a restoring torque which tends to return the longitudinal axis to alignment with the local vertical.

The difference between the gravitational and the centrifugal force acting on either mass of the dumbbell can be resolved along and perpendicular to the dumbbell axis. The perpendicular component results in a torque about the center of mass while the component along the dumbbell provides a tension force which can maintain structural stability even for nonrigid bodies such as tether.

Major hardware components for passive gravity gradient stabilization are booms and dampers. The booms and their tip masses can be deployed in orbit to produce required inertias. Damping in the form of eddy currents or hysteresis losses is added to the vehicles so that any librations about the local vertical are reduced in magnitude. The dampers are anchored to some independent reference such as the Earth's magnetic or gravitational field and the rotational energy is dissipated when the main body of the satellite moves relative to the damper body.

7.1.1 Booms for Gravity Gradient Systems

Long space-erectable booms extending from the spacecraft structure provide gravity gradient stabilized satellites with moments of inertia large enough to generate high stabilizing torques. A common type of boom used for this purpose is a flat tape. Upon emerging from a reel dispenser, as illustrated in Figure 7.2, it forms a tube of circular cross section. A boom of this type, however, has been found to respond to incident solar thermal energy with a degree of bending, which can cause significant impairment of pointing accuracy.

In space, booms undergo two distinct modes of thermal bending: (1) dynamic thermal bending or a "snapping" of the boom as the satellite crosses from the Earth's shadow into sunlight, which induces a high-frequency oscillation in the satellite-boom combination, and (2) static thermal bending, which is the steady-state offset of the boom on which the sun shines.

The bending of the boom changes the angle between the satellite's vertical axis and the local vertical. The magnitude of this effect depends on the parameters of the satellite and its orbit. For example, the error for a boom 30 meters long

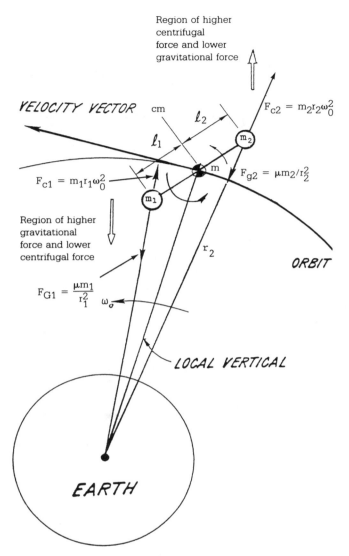

Figure 7.1 Gravity gradient restoring torque.

provides a much higher torsional and bending resistance capability for an improved load-carrying capacity.

7.1.2 Libration Dampers

Several types of libration dampers are used for reducing the oscillatory motion of gravity gradient stabilized systems. A magnetic fluid damper, for example, is illustrated in Figure 7.4(a). The damper consists of an inner sphere, containing a strong permanent magnet, and an outer sphere, lined internally with a shell of pyrolytic graphite. The inner sphere is suspended in the outer sphere by the repulsive forces between the permanent magnet and the graphite shell. A gap, filled with a viscous fluid, is provided between the inner and the outer spheres. Because of the permanent magnet, the inner sphere will track the Earth's magnetic field, while the outer sphere attached to the satellite through the extendable rods will be subjected to the gravity gradient torque. Since the inner sphere can rotate with respect to the outer sphere, a difference in angular velocity between the inner and outer spheres will result because of satellite librations. This relative angular velocity will produce a shearing action on the viscous fluid, and damping will result through the dissipation of energy.

Another type of libration damping is illustrated in Figure 7.4(b). The time-lag magnetic damping method uses the interaction between the Earth's magnetic field and the satellite's magnetic dipole generated by three orthogonal electromagnets (coils). The magnetic field is sensed in the spacecraft coordinates and stored in a computer memory for a specified period of time. The electromagnets are energized to produce a magnetic dipole which produces a torque on the spacecraft to oppose the librational motion of the satellite. A variant of this approach is a single coil (e.g., a z-axis coil) which is oriented 90° to the magnetic field. A pitch torque on the spacecraft may thus be produced to invert a gravity gradient stabilized satellite, if necessary.

An example of a two axis (dumbbell) stabilized spacecraft is shown in Figure 7.5(a).

The UOSAt is named for the University of Surrey, England, where it was constructed. UOSAts 1 and 2 were launched in October 1981 and March 1984. Both are in polar orbits, the first at about 500 kilometers altitude, the second at about 700 kilometers.

Some aspects of the theory and flight experience of gravity gradient stabilized systems are discussed in references 1 through 3. Other applications of gravity gradient stabilization are described in references 4 through 10.

NASA's Long Duration Exposure Facility (LDEF), another example of a gravity gradient stabilized satellite, is illustrated in Figure 7.5(b). The cylindrically shaped satellite was deployed in a low altitude (approximately 500 km) circular orbit by the Space Shuttle and was recovered from orbit nearly six years later in January of 1990.

made of 0.005 centimeter thick beryllium-copper tape could be on the order of 10° due to static bending and a similar amount due to dynamic bending. The outer surface of the boom can be silver plated or perforated to reduce these effects somewhat. An example of the latter approach is an "unbendable" boom developed for the Radio Astronomy Explorer (RAE-1) satellite, launched on 4 July 1968, which contained a pattern of holes in combination with selected thermal coatings on the inner and outer surfaces of the boom. The extended length of that boom was 457.0 meters. To eliminate the low torsional rigidity of the open section booms, the boom had an interlocked seam which enhanced its torsional rigidity. See Figure 7.2.

A telescoping boom design using a combination of hinges and box-like sections is illustrated in Figure 7.3. This design

Overlap

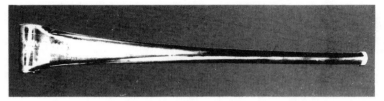

Edgelock

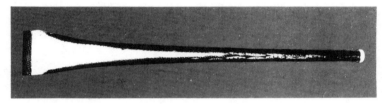

Hingelock

(a) Motorized
     Erection Unit

Figure 7.2 "Fairchild" type tubular elements (booms).

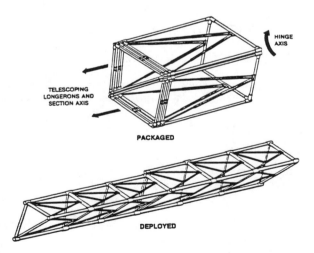

Figure 7.3 Telescoping boom. (Courtesy of NASA Tech Briefs, Spring 1985).

7.2 *Equations of Motion*

For a satellite with orbital angular velocity ω_s and principal moments of inertia $I_\alpha(\alpha = 1, 2, 3)$ in a circular orbit, Euler's equation of motion in terms of $\vec{h}$, the spacecraft an-

gular momentum vector, and $\vec{T}^{(g)}$, the gravity gradient torque vector, is

$$\dot{\vec{h}} + \vec{\omega}_s \times \vec{h} = \vec{T}^{(g)} \tag{7.1}$$

where

$$\vec{h} = \omega_1 I_1 \hat{e}_1 + \omega_2 I_2 \hat{e}_2 + \omega_2 I_3 \hat{e}_3$$
$$\vec{T}^{(g)} = T_1^{(g)} \hat{e}_1 + T_2^{(g)} \hat{e}_2 + T_3^{(g)} \hat{e}_3$$

Referring to Figure 7.6 the orientation of the spacecraft principal axes, coinciding with the unit vectors $\hat{e}_\alpha$, is related to the orbiting reference frame $\hat{E}_\beta$ through three sequential rotations θ_1, θ_2, and θ_3. The body attitude matrix can be written as in equation 1.9

$$\begin{pmatrix} \hat{e}_1 \\ \hat{e}_2 \\ \hat{e}_3 \end{pmatrix} = \tag{7.2}$$

$$\begin{pmatrix} c\theta_3 & s\theta_3 & 0 \\ -s\theta_3 & c\theta_3 & 0 \\ 0 & 0 & 1 \end{pmatrix} \begin{pmatrix} c\theta_2 & 0 & -s\theta_2 \\ 0 & 1 & 0 \\ s\theta_2 & 0 & c\theta_2 \end{pmatrix} \begin{pmatrix} 1 & 0 & 0 \\ 0 & c\theta_1 & s\theta_1 \\ 0 & -s\theta_1 & c\theta_1 \end{pmatrix} \begin{pmatrix} \hat{E}_1 \\ \hat{E}_2 \\ \hat{E}_3 \end{pmatrix}$$

$$= \begin{pmatrix} (c\theta_2 c\theta_3) & (c\theta_1 s\theta_3 + c\theta_3 s\theta_1 s\theta_2) & (-c\theta_3 s\theta_2 c\theta_1 + s\theta_3 s\theta_1) \\ (-c\theta_2 s\theta_3) & (c\theta_1 c\theta_3 - s\theta_1 s\theta_2 s\theta_3) & (s\theta_3 s\theta_2 c\theta_1 + c\theta_3 s\theta_1) \\ (s\theta_2) & (-c\theta_2 s\theta_1) & (c\theta_2 c\theta_1) \end{pmatrix} \begin{pmatrix} \hat{E}_1 \\ \hat{E}_2 \\ \hat{E}_3 \end{pmatrix}$$

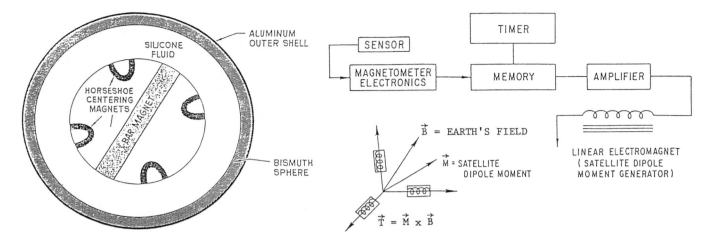

(a) Magnetically anchored viscous fluid damper.   (b) Time-lag magnetic damping scheme.

Figure 7.4 Libration dampers.

The body components of the spacecraft angular velocity $\vec{\omega}_s$ are given by

$$\omega_\alpha = \dot{\theta}_\alpha + \omega_o \hat{E}_2 \cdot \hat{e}_\alpha \qquad (7.3)$$

or in expanded form as

$$\begin{aligned}
\omega_1 &= \dot{\theta}_1 + \omega_o(c\theta_3 s\theta_1 s\theta_2 + c\theta_1 s\theta_3) \\
\omega_2 &= \dot{\theta}_2 + \omega_o(c\theta_1 c\theta_3 - s\theta_1 s\theta_2 s\theta_3) \\
\omega_3 &= \dot{\theta}_3 + \omega_o(-s\theta_1 c\theta_2)
\end{aligned} \qquad (7.4)$$

where $\dot{\theta}_\alpha(\alpha = 1-3)$ are the angular rates about the sequential rotation axes. Linearizing yields

$$\begin{aligned}
\omega_1 &= \dot{\theta}_1 + \omega_o\theta_3 \\
\omega_2 &= \dot{\theta}_2 + \omega_o \\
\omega_3 &= \dot{\theta}_3 - \omega_o\theta_1 .
\end{aligned} \qquad (7.5)$$

The component equations in scalar form are

$$\begin{aligned}
I_1\dot{\omega}_1 + \omega_2\omega_3(I_3 - I_2) &= T_1^{(g)} \\
I_2\dot{\omega}_2 + \omega_1\omega_3(I_1 - I_3) &= T_2^{(g)} \\
I_3\dot{\omega}_3 + \omega_1\omega_2(I_2 - I_1) &= T_3^{(g)}
\end{aligned} \qquad (7.6)$$

which, upon substitution of the linearized expressions for ω_α, become

$$I_1(\ddot{\theta}_1 + \omega_o\dot{\theta}_3) + (\dot{\theta}_2 + \omega_o)(\dot{\theta}_3 - \omega_o\theta_1)(I_3 - I_2) = T_1^{(g)}$$

$$I_2\ddot{\theta}_2 + (\dot{\theta}_1 + \omega_o\theta_3)(\dot{\theta}_3 - \omega_o\theta_1)(I_1 - I_3) = T_2^{(g)}$$

$$I_3(\ddot{\theta}_3 - \omega_o\dot{\theta}_1) + (\dot{\theta}_1 + \omega_o\theta_3)(\dot{\theta}_2 + \omega_o)(I_2 - I_1) = T_3^{(g)} \quad (7.7)$$

$T_1^{(g)}$, $T_2^{(g)}$, and $T_3^{(g)}$ are the body components of the gravity gradient torque. For a circular orbit they are of the form

$$\begin{pmatrix} T_1^{(g)} \\ T_2^{(g)} \\ T_3^{(g)} \end{pmatrix} = 3\omega_o^2 \begin{pmatrix} (I_3 - I_2) & a_{21}a_{31} \\ (I_1 - I_3) & a_{11}a_{31} \\ (I_2 - I_1) & a_{11}a_{21} \end{pmatrix} . \qquad (7.8)$$

For small angular deviations θ_1, θ_2, and θ_3, the torque components can be expressed by the following approximation

$$\begin{pmatrix} T_1^{(g)} \\ T_2^{(g)} \\ T_3^{(g)} \end{pmatrix} \approx 3\omega_o^2 \begin{pmatrix} 0 \\ (I_1 - I_3)\theta_2 \\ (I_1 - I_2)\theta_3 \end{pmatrix} . \qquad (7.9)$$

Substituting these torques into equations 7.7, the linearized equations become

$$I_1(\ddot{\theta}_1 + \omega_o\dot{\theta}_3) + (I_2 - I_3)(\omega_o^2\theta_1 - \omega_o\dot{\theta}_3) = 0$$

$$I_2\ddot{\theta}_2 + 3\omega_o^2(I_3 - I_1)\theta_2 = 0$$

$$I_3(\ddot{\theta}_3 - \omega_o\dot{\theta}_1) + (I_2 - I_1)(4\omega_o^2\theta_3 + \omega_o\dot{\theta}_1) = 0 . \qquad (7.10)$$

7.3 Restoring Torques

The approximate control (restoring) torques about the e_α body fixed axes for small deviations from the reference frame can be obtained by neglecting the small $\dot{\theta}$ coupling terms in equation 7.10. Thus, the linearized restoring torques are of the form

$$\begin{aligned}
T_1 &= -\omega_o^2(I_2 - I_3)\theta_1 \\
T_2 &= -3\omega_o^2(I_3 - I_1)\theta_2 \\
T_3 &= -4\omega_o^2(I_2 - I_1)\theta_3
\end{aligned} \qquad (7.11)$$

It can be seen that the restoring torques vanish when the principal moments of inertia are identical.

7.4 Natural Frequencies

The pitch (θ_2) equation, which is uncoupled from the roll (θ_3) and yaw (θ_1) equations, is given as

$$\ddot{\theta}_2 + \frac{3\omega_o^2(I_3 - I_1)\theta_2}{I_2} = 0 \qquad (7.12)$$

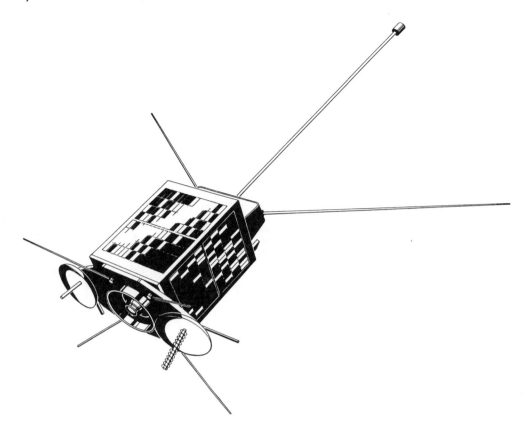

Figure 7.5 Examples of gravity gradient stabilized satellites: (a) UOSat and (b) the Long Duration Exposure Facility (LDEF).

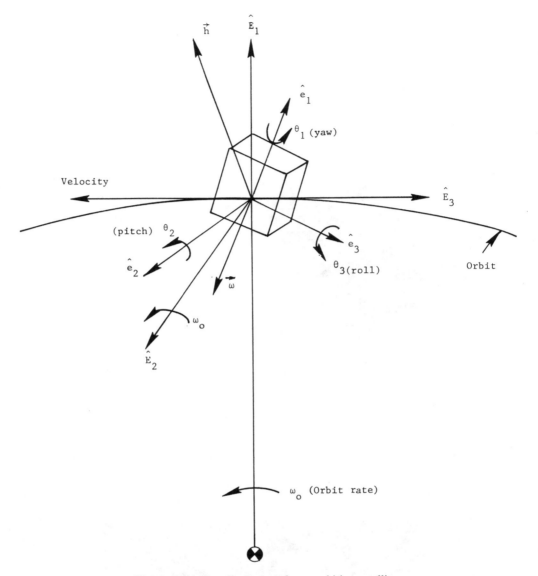

Figure 7.6 Coordinate axes for an orbiting satellite.

or

$$\ddot{\theta}_2 + \omega_2^2\theta_2 = 0 \qquad (7.13)$$

where

$$\omega_2^2 = \frac{\omega_o^2(I_3 - I_1)}{I_2}.$$

Similarly, the roll and yaw natural frequencies for small coupling can be approximated as

$$\omega_3^2 = \frac{4\omega_o^2(I_2 - I_1)}{I_3} \qquad (7.14)$$

and

$$\omega_1^2 = \frac{\omega_o^2(I_2 - I_3)}{I_1}.$$

For a dumbbell satellite $I_1 \approx 0$ and $I_2 = I_3$ and the libration frequencies become

$$\omega_2^2 = 3\omega_o^2$$
$$\omega_3^2 = 4\omega_o^2$$
$$\omega_1^2 = 0 \qquad (7.15)$$

For a planar body $I_2 = I_1 + I_3$, and the frequencies become

$$\omega_2^2 = 3\omega_o^2(2I_3/I_2 - 1)$$
$$\omega_3^2 = 4\omega_o^2$$
$$\omega_1^2 = \omega_o^2 \qquad (7.16)$$

which indicates a resonance in yaw and roll motions with the orbital rate ω_o.

The natural frequency map is shown in Figure 7.7. For example, the point at $I_3/I_2 = 1$ and $I_1/I_2 = 0$ represents a

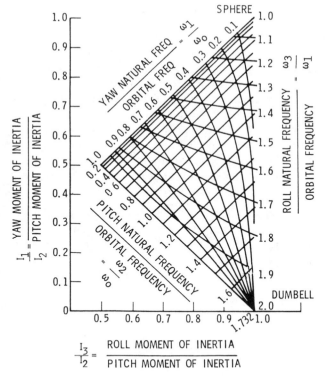

Figure 7.7 Natural frequency map.

theoretical dumbbell which has natural oscillation (libration) frequencies of $1.732\ \omega_o$ in pitch (in the orbit plane) and $2\ \omega_o$ in roll (out of the orbit plane). There is no oscillation about the local vertical since this motion is uncontrolled; that is, there is zero yaw restoring torque. Also, large amplitude oscillations can be produced in roll by relatively small disturbance torques (arising from solar radiation pressure, for example) at the $2\ \omega_o$ frequencies. It is generally desirable to select moments of inertia which avoid such resonance conditions. The plot in Figure 7.7 was generated numerically from the solution of the coupled roll/yaw equations and the uncoupled pitch equation [2].

7.5 Stability Considerations

In general, there are two stable spacecraft orientations for the case of a circular orbit. The conventional stability configuration is shown in Figure 7.8a. In this case the axis of minimum moment of inertia is aligned with the radial direction and the axis of maximum moment of inertia is normal to the orbit plane; that is

$$I_1 < I_3 < I_2 \tag{7.17}$$

where I_1, I_2, and I_3 are the yaw, pitch, and roll moments of inertia, respectively. For this case, the rotation of a gravity gradient stabilized body takes place at orbital rate ω_o about

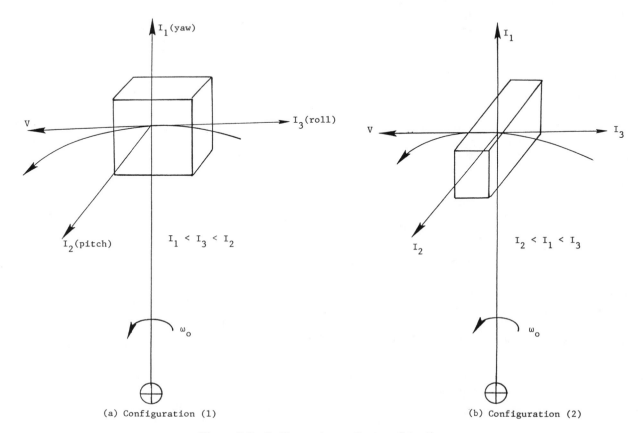

Figure 7.8 Stable gravity gradient configurations.

the major principal axis (the pitch axis $\hat{e}_2$) which implies stability with or without damping or energy dissipation in the system, since the potential energy is zero in the nominal orientation.

For the second configuration illustrated in Figure 7.8b, the principal axis of intermediate moment of inertia is aligned with the local vertical, and the axis of minimum moment of inertia is normal to the orbit plane. For this case, the motion is stable as a result of gyroscopic coupling. The potential energy in the nominal orientation (zero restoring torque position) is a "saddle" point. The motion about the pitch axis increases the potential energy but rotation about the roll axis decreases it. Thus, the internal energy dissipation causes instability just as a gyro or a body rotating about its axis of minimum moment of inertia (in the absence of external torques) drifts away from its equilibrium position due to energy dissipation. By contrast with the conventional configuration, the oscillations tend to diverge from the initial conditions until the angular velocity has increased sufficiently to supply the gyroscopic torques which stabilize the body.

7.6 Eccentric Orbit

For a slightly eccentric orbit ($\varepsilon \neq 0$), the equations of motion are qualitatively different from those for a circular orbit in that their coefficients are variable (periodic). For example, consider the pitch motion in an elliptic orbit with $\omega_2 \approx \dot{\theta}_2 + \omega_o(1 + 2\varepsilon \cos \omega_o t)$. The pitch equation then becomes

$$I_P \dot{\omega}_2 = T_2 \qquad (7.18)$$

or approximately

$$I_2(\ddot{\theta}_2 - 2\omega_o^2\varepsilon \sin \omega_o t) = \frac{3\mu}{R_o^3}(I_1 - I_3)a_{11}a_{31} \qquad (7.19)$$

$$= 3\omega_o^2(1 + 3\varepsilon \cos \omega_o t)(I_1 - I_3)\theta_2 .$$

This can be written as

$$\ddot{\theta}_2 + 3\omega_o^2(1 + 3\varepsilon \cos \omega_o t)\frac{(I_3 - I_1)}{I_2}\theta_2$$
$$= 2\omega_o^2\varepsilon \sin \omega_o t \qquad (7.20)$$

which shows both forced and parametric excitation due to orbit eccentricity ε. Thus, satellites which have a stable attitude when in a circular orbit are subject to excitations which can make them tumble when the orbit is eccentric. The orbital eccentricity, above which tumbling occurs, varies with the vehicle shape but is generally between 0.05 and 0.2 for most vehicles.

When gravity external torques alone are considered, linearized equations with constant coefficients make a good mathematical model of the system as long as the angles stay small and there is no chance of instability developing.

7.7 Capture Requirements

Gravity gradient stabilized satellites are inherently bi-stable on all axes. For missions where a particular face of the satellite must point toward the Earth, capture in the right-side-up orientation must be provided for. This can be done either by deploying gravity gradient rods immediately upon separation from the stabilized booster (so as to achieve capture in the correct orientation), or by including turnover capability in the satellite with a reaction wheel or a magnetic torquer. For many applications, the capture requirements may be one of the strongest factors dictating the moment of inertia of the deployed system.

The satellite will be captured whenever its potential and kinetic energies together are less than a "threshhold" capture energy. This threshold capture energy E_t corresponds to the energy of the satellite oscillating between ± 90 degrees, just short of tumbling. Thus, in pitch

$$E_t = \int_o^{\pi/2} \text{(gravity torque) } d\theta_2$$

$$= \int_o^{\pi/2} \left(\frac{K^{(g)}}{2} \sin 2\theta_2\right) d\theta_2 = \frac{K^{(g)}}{2} \qquad (7.21)$$

where $K^{(g)}$ is the gravity gradient constant. Also, the potential energy E_p is given by

$$E_P = \int_0^{\theta_2} \text{(gravity torque) } d\theta$$

$$= \int_0^{\theta_2} \left(\frac{K^{(g)}}{2} \sin 2\theta\right) d\theta$$

$$= -\frac{K^{(g)}}{4} (\cos 2\theta_2 - 1) \qquad (7.22)$$

and the kinetic energy E_k by

$$E_k = \frac{1}{2} I_2 \dot{\theta}_2^2 . \qquad (7.23)$$

The threshold capture energy E_t is

$$E_t = E_p + E_k$$

or

$$\frac{1}{2} K^{(g)} = \frac{1}{4} K^{(g)}(1 - \cos 2\theta_2) + \frac{1}{2} I_2 \dot{\theta}_2^2$$

which yields

$$\dot{\theta}_2 = \sqrt{\frac{K^{(g)}}{I_2}} \cos \theta_2 \qquad (7.24)$$

as the critical pitch angular rate for capture. Similarly, the gravity gradient constants are

$$K_2^{(g)} = 3\omega_o^2(I_3 - I_1) \quad \text{for pitch}$$
$$K_3^{(g)} = 4\omega_o^2(I_2 - I_1) \quad \text{for roll}$$

and

$$K_1^{(g)} = \omega_o^2(I_2 - I_3) \quad \text{for yaw .} \quad (7.25)$$

The critical angular rates for capture become

$$\dot\theta_2 = \sqrt{3}\omega_o \sqrt{\frac{I_3 - I_1}{I_2}} \cos\theta_2$$

$$\dot\theta_3 = 2\omega_o \sqrt{\frac{I_2 - I_1}{I_3}} \cos\theta_3$$

$$\dot\theta_1 = \omega_o \sqrt{\frac{I_2 - I_3}{I_1}} \cos\theta_1 \quad (7.26)$$

7.7 Steady State Requirements

Selection of the "best" values to use for the three principal moments of inertia requires an optimization with regard to the specific mission constraints. If the pitch inertia requirement is dictated by capture considerations, then only the selection of roll and yaw inertias remains. The natural frequency map in Figure 7.7 permits the determination of the natural frequencies of pitch, roll, and yaw as a function of the moment of inertia ratios. As a rule, the pitch and yaw natural frequencies are the determining frequencies and are selected to provide good frequency response and large restoring torques. Since most of the disturbance torques occur at the orbital frequency, this point must be avoided (on pitch, roll, and yaw), and the moment of inertia ratios of interest are in the lower portion of the map. The particular values of moments of inertias to be used for a given application depend upon the particular vehicle and orbit parameters and the specific performance requirements for that mission.

Further reduction in pointing errors can be achieved by reducing the disturbance torques and increasing the restoring torques. Both must be pursued, but to attain minimum weight it is desirable to use a system having the lowest moments of inertia that will meet performance requirements. Minimization of the disturbance torques is, therefore, the major approach. To achieve this, an understanding of the nature and importance of the individual disturbances is necessary.

7.8 Space Shuttle Gravity Gradient Stabilization—An Example

The stable equilibrium orientation of the Shuttle in orbit is shown schematically in Figure 7.9 (wings in the orbit plane, nose pointing away from Earth). The body fixed axes e_1, e_2, and e_3 (corresponding to the vehicle principal axes) are shown aligned with an Earth-following orbiting frame E_1, E_2, and E_3, where E_1 is along the outward radius vector,

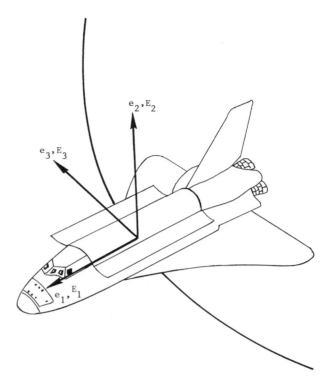

Figure 7.9 Space Shuttle body axes.

E_3, is along the negative orbital velocity vector and E_2 is normal to the orbit plane. This equilibrium orientation requires that the moment of inertia condition in equation 7.17 is satisfied, that is,

$$I_1 < I_3 < I_2 \quad (7.27)$$

where I_1, I_2, and I_3 are the yaw, pitch, and roll principal moments of inertia, respectively.

The degree of resistance to external disturbances provided by the gravity gradient effect can be evaluated in terms of the gravitational restoring torques about the orbiting reference frame. Thus, the gravity gradient restoring torques $T_\alpha^{(g)}(\alpha = 1 - 3)$ given by equation 7.11 are repeated here as

$$T_1 = -\omega_o^2(I_2 - I_3)\theta_1$$
$$T_2 = -3\omega_o^2(I_3 - I_1)\theta_2$$
$$T_3 = -4\omega_o^2(I_2 - I_1)\theta_3 . \quad (7.28)$$

For a 278 kilometer altitude circular orbit

$$T_1/\theta_1 = -0.01 \text{ N} \cdot \text{m/deg}$$
$$T_2/\theta_2 = -0.56 \text{ N} \cdot \text{m/deg}$$
$$T_3/\theta_3 = -0.79 \text{ N} \cdot \text{m/deg} \quad (7.29)$$

The corresponding libration (oscillation) frequencies are written approximately as

$$\omega_1 \approx 0.5\,\omega_o, \quad \omega_2 \approx 1.6\,\omega_o, \quad \omega_3 \approx 1.92\,\omega_o \quad (7.30)$$

indicating that the yaw period ($P_1 = 2\pi/\omega_1$) is very long (about two orbital revolutions) and that the pitch (P_2) and roll (P_3) periods are somewhat shorter. The librational motions are therefore predictable, and to some extent, controllable, as well as possibly beneficial from a thermal balance point of view.

7.8.1 Aerodynamic Torque Effects

The presence of aerodynamic torques at altitudes of less than 600 kilometers affects the equilibrium orientation of the vehicle as well as the amplitude and frequency of the resultant oscillations. New equilibrium orientations can be found by solving the equation

$$\dot{\vec{H}} = \vec{T}^{(g)} + \vec{T}^{(a)} - \vec{\omega}_o \times \vec{H} \qquad (7.31)$$

where $\vec{T}^{(g)}$ and $\vec{T}^{(a)}$ are the gravity and aerodynamic torques, respectively.

7.8.2 Flight Results

During the STS-4 mission a test of gravity gradient stabilization was performed. Figure 7.10 displays oscillations of the Shuttle Orbiter as a function of mission time. Oscillations about the pitch and roll axes were within a few degrees of the nominal orientation. The oscillation about the yaw axis (along the local vertical) was significantly greater ($\sim \pm 20°$) as the result of a very low restoring torque about this axis ($T_1/\theta_1 \approx -0.01$ N $\cdot$ m/deg).

7.9 Tethered Satellite Systems

The idea of using very long structures in space is not new (see references 11–16, for example). The earliest suggestion of such use is credited to Tsiolkovsky [13] who visualized in 1895 a tower that would extend from the equator of the Earth to geosynchronous altitude and beyond. Later, interest in tethered systems was initially associated with similar futuristic schemes, as in references 4 through 16, for example, or with the thought of retrieving astronauts stranded in space. The feasibility of using tethers was established during two Gemini missions when the manned Gemini XI space vehicle was linked by tether to an unmanned Agena vehicle [11].

7.9.1 Tethered Satellite System Experiment

The planned tethered satellite system (TSS) experiment consists of the Space Shuttle mounted deployer and a tether supported satellite which is to be deployed downward toward Earth or upward away from Earth to maximum distances of 100 kilometers downward or 20 kilometers upward. The primary purpose of the experiment is to develop a capability

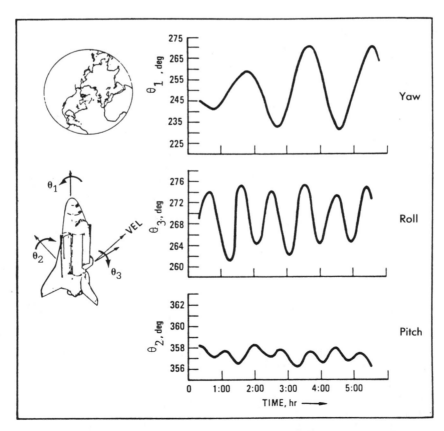

Figure 7.10   Space Shuttle attitude excursions in gravity gradient mode in an orbital coordinate frame.

to perform a variety of scientific investigations to study geomagnetism, electrodynamic effects, atmospheric properties, and chemical release effects. Figure 7.11 illustrates the TSS configuration with a coordinate system at the origin of the Space Shuttle center of gravity.

7.9.1.1 Deployment and Retrieval

Tether tension and boost force F (gravity gradient, Coriolis, thruster, etc.) are used in a closed loop control system to deploy a subsatellite (probe) at the end of the tether toward or away from the Earth. A control law has been studied in reference 18 which is of the form

$$F = K_1\ell + C_1\dot{\ell} + K_2\ell_c \qquad (7.32)$$

where ℓ and $\dot{\ell}$ are instantaneous length and length rate, respectively, ℓ_c is a commanded length, while K_1, K_2, and C_1 are a set of constants. ℓ_c is changed in steps until the final tether length is attained. Implementation of the control law required the measurement of tether tension, rate of tether deployment, and instantaneous tether length. The latter two can be measured by a pulley kept in frictional contact with the tether. The tension can be measured by a spring damper arrangement on the same pulley. The measurements are fed into a computer which calculates the required torque that must be produced by the motor driving the tether reel system.

This control law was found to be effective for damping in-plane motion during deployment and retrieval. The C_1 coefficient damps the longitudinal oscillations of the tether. Out-of-plane motion was not considered in the analysis.

A somewhat improved deployment control system was considered by Kissel et al., [19] with a modified tension law and a commanded length of the form

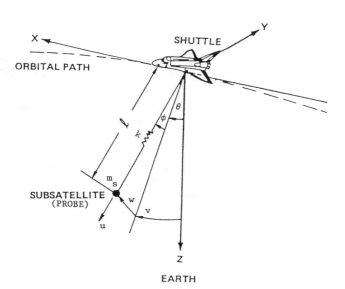

Figure 7.11   Tethered subsatellite coordinate system.

$$\ell_c = K_1\ell + K_3 \qquad (7.33)$$

where K_3 is a constant.

Retrieval in two phases is considered in reference 20. Also, a deployment without active control is discussed in reference 21, where a passive procedure is initiated by placing the tether (except its end points) outside the Shuttle and allowing it to float freely. Next, the payload attached to one end of the tether is ejected from the Shuttle and it performs a free flight until the tether becomes taut, at which time the payload is subjected to an impulse affecting the motion. Impacts and free flights occur alternately until enough energy has been dissipated during impacts so that no further ones occur. The tether then becomes permanently taunt and the system behaves like a spherical pendulum. If some viscous damping is provided, the pendulum settles along the local vertical. Deployment direction (upward or downward), and the time required to deploy, depend on the initial ejection velocity.

7.9.1.2 Steady-State Tension and Elongation

The steady-state tension along a massless boom of length ℓ connecting two masses m_1 and m_2, as illustrated in Figure 7.1, is of the form

$$F = 3\omega_o^2\ell m \qquad (7.34)$$

where ω_o is the angular rate of the circular orbit, $\ell = \ell_1 + \ell_2$ and $m = m_1 m_2/(m_1 + m_2)$ is the reduced mass of the system. The tension is the difference between the centrifugal and gravitational forces at each mass resolved along the booms of length ℓ_1 and ℓ_2 measured from the center of gravity to each mass, respectively.

With respect to Figure 7.11 the tension along the length of the tether can be expressed in the form [11]

$$F = 3\,m_s\omega_o^2\ell[1 + (r/2)(1 - X^2)] \qquad (7.35)$$

where ℓ, m_s, and ω_o are tether length, subsatellite mass, and orbital rate, respectively; r is the mass ratio m_t/m_s, where m_t is the tether mass. Here, X is the dimensionless distance of a given point from the Shuttle. Equation 7.34 is an approximation of the actual cosine variations of the tension presented by Misra and Modi [28] and Bergamaschi [27]. It ignores the contribution of the tether mass to the tension. The static elongation is given by

$$v_s = (3\,m_s\omega_o^2\ell^2/EA)[X + (r/2)(X - X^3/3)] \qquad (7.36)$$

where E is the modulus of elasticity and A the cross sectional area of the tether. At the probe end ($X = 1$), the static elongation is

$$v_s = (3\,m_s\omega_o^2\ell^2/EA)[1 + (r/3)] \qquad (7.37)$$

which is of the order of 100 meters for the second TSS mission. More accurate sine variations of the elongation are given in references 27 and 28.

7.9.1.3 Elastic Oscillations of Tethers

The frequencies of longitudinal oscillations are given in reference 11 as

$$\omega_{\ell_n} = \lambda_n \omega_o \qquad (7.38)$$

where λ_n are the roots of the transcendental equation

$$\tan\left[(3 + \lambda^2)^{1/2}/p\right] = rp/(3 + \lambda^2)^{1/2} \qquad (7.39)$$

where

$$p^2 = EA/\rho\omega_o^2\ell^2$$

and ρ is the mass density of the tether. It can be shown from equation 7.38 that the lowest frequency ω_{ℓ_1} is approximately given by

$$\omega_{\ell_1} = [EA/m_s\ell]^{1/2} \qquad (7.40)$$

while the higher frequencies of longitudinal oscillations can be obtained roughly by using

$$\omega_{\ell_n} = [(n-1)\pi/\ell][EA/p]^{1/2} , \quad n = 2, 3, \ldots \qquad (7.41)$$

A plot of the first two longitudinal oscillation natural frequencies are shown in Figure 7.12 as a function of tether length.

A somewhat different approach is taken in reference 17. Because of the underlying assumptions of the analysis, the only structural potential energy taken into account is the strain energy related to the axial extension of the tether. In the case that the tether can be assumed to be uniformly deformed, i.e., the displacement of any reference point on the undeformed tether is proportional to its distance from the attachment point x, the strain energy of the tether takes the same form as the potential energy of a massless spring; that is

$$\text{P. E.} = \frac{1}{2} k(x - \ell)^2$$

where

$$k = AE/\ell \qquad (7.42)$$

Here, A and ℓ are the cross-sectional area and length of the undeformed tether, and E is the tensional elasticity modulus of the tether material, as before. The frequency of the axial oscillations of the free mass-spring system is given by

$$\omega_\ell = [k/(\bar{m} - m_t/6)]^{1/2} \qquad \text{for } x > \ell$$

and

$$\omega_o = 0 \qquad \qquad \text{for } x < \ell . \qquad (7.43)$$

Here, $\bar{m}$, the reduced mass of the tether system, is

$$\bar{m} = (m_s + m_t/2)(m_o + m_t/2)/m \qquad (7.44)$$

m_o, m_s, and m_t represent the masses of the Shuttle, subsatellite (probe), and tether, respectively, and $m = m_o + m_s + m_t$ is the sum of these masses. The parameter κ denotes the mass ratio of the two lumped masses. For a typical tether the mass density is of the order of 1 kilogram per kilometer so that the tether mass would usually be less than the end masses.

For typical tether materials (e.g., Aramid polymers) with a longitudinal stiffness AE in the neighborhood of 10^4 N and a typical tether configuration with ℓ in the range of 10 to 100 kilometers and $\bar{m}$ from 100 to 1000 kilograms, the tether's axial frequency ranges from about 10 to 100 times the orbital frequency. As can be seen from equations 7.42 and 7.43, the frequency increases with decreasing tether length and decreasing reduced mass.

The transverse vibrations of the tether are presented in Table 7.2, and the corresponding mode shapes for two natural frequencies are found in Figure 7.13 [20]. The natural frequencies in Table 7.2 are normalized to the mean orbit rate ω_o. The v and w subscripts denote vibrations in the orbital plane and normal to it, respectively (see Figure 7.11). The results show that the first natural frequency (mode shape) corresponds to the rigid body rotation of the tether supported at the origin of the reference frame. The second mode represents the first flexible mode of the tether supported at the origin and free at the probe end. The deflection at the probe end is a function of the mass ratio m_s/m_t.

The natural frequencies of motion in the orbital plane can be approximately expressed as [11]

$$\omega_{t_n} = k_n[F/\rho\ell^2]^{1/2} , \qquad n = 1, 2, \ldots \qquad (7.45)$$

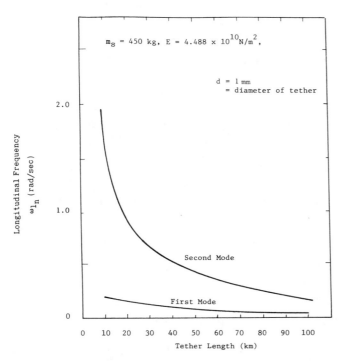

Figure 7.12 Variation of frequencies of longitudinal oscillations with the length of the tether [11].

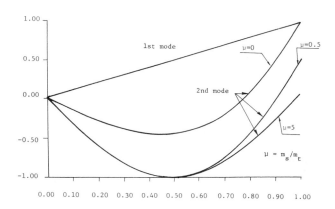

Figure 7.13. Selected mode shapes for the natural frequencies in Table 7.2 [20].

where F is the steady tension at the probe end. Using equation 7.34, this can be rewritten as

$$\omega_{t_n} = k_n (3/r)^{1/2} \omega_o = k_n (3m_s/m_t)^{1/2} \omega_o \qquad (7.46)$$

where

$$k_1^2 = \pi^2 (1 + r/3) - (r/4)$$
$$k_2^2 = 4\pi^2 (1 + r/3) - (r/4)$$

Here

ω_o = orbital angular frequency
$r = m_t/m_s$ is the mass ratio
ρ = tether density.

Thus, the transverse vibrational frequencies are linearly dependent on the orbital frequency ω_o and increase with the probe mass. If the tension in the tether were only due to the probe, and hence constant along the length, k_n would have been equal to $n\pi$.

The $n = 1$ case in equation 7.45 is the first flexible mode corresponding to mode two in Table 7.2. Free-free boundary conditions or end constraints are considered in reference 23. Relative motion and tether elastic vibrations are also investigated in references 24 through 28.

7.9.2 Tether Propulsion

Tethers have a great potential for raising payloads to higher orbits (tether propulsion [29]). Tethered deployment of two masses in Earth orbit is based on the transfer of angular

momentum from one mass to the other, while the total angular momentum is conserved. The possible applications are (1) tethered launch of an Orbiter payload to any higher orbit including geosynchronous orbit; (2) tethered deployment of an Orbiter from the Space Station; (3) tethered launch of an orbital transver vehicle (OTV) from Space Station orbits, etc. An illustration of a tethered Space Station-Shuttle configuration is shown in Figure 7.14.

7.10 Recommended Practice for Gravitational Stabilization of Satellites

The following general considerations apply to the design of passively stabilized satellites.

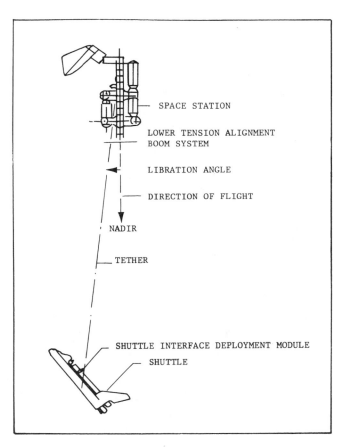

Figure 7.14 Space Station/Shuttle tethered system.

Table 7.2 Dimensionless Natural Frequencies for Shuttle Supported Probe on Tether

Mode	$m_s/m_t = 0$		$m_s/m_t = 0.5$		$m_s/m_t = 1$		$m_s/m_t = 5$	
	ω_v/ω_o	ω_w/ω_o	ω_v/ω_o	ω_w/ω_o	ω_v/ω_o	ω_w/ω_o	ω_v/ω_o	ω_w/ω_o
$n = 1$	1.732	2.000	1.732	2.000	1.732	2.000	1.732	2.000
$n = 2$	4.243	4.359	5.515	5.5605	6.707	6.782	12.776	12.815
$n = 3$	6.708	6.782	10.134	10.1825	12.742	12.782	25.174	25.193

a. Select inertia augmentation booms and end masses to provide gravity gradient restoring torques consistent with specified pointing requirements. The restoring torques should be at least several times the maximum environmental disturbance torques.

b. For the best performance, consider the following:
   1. Minimum number of booms (preferably one) as short as possible (e.g., 10 to 15 meters) to minimize thermal and other environmental effects.
   2. Polished silver-plated exterior booms resistant to thermal deformation and flutter.
   3. Boom motion damper as part of the end mass.
   4. Gravitational and aerodynamic effects on boom bending where applicable.
   5. Natural bending frequency and structural damping of booms to avoid interaction with any active elements (e.g., reaction wheels of thrusters) in the spacecraft.

c. Provide for reorienting the spacecraft should the stable vehicle orientation be reversed due to disturbance torques. A small pitch reaction wheel can be used for this purpose.

7.11 References

1. Hughes, P. C. *Spacecraft Attitude Dynamics*, John Wiley and Sons, 1986.
2. DeBra, D. B., and R. H. Delp. "Rigid Body Attitude Stability and Natural Frequencies in a Circular Orbits," J. Astron., Sci., 8(1): 14–17 (Spring 1961).
3. Williamson, R. K. "Some Aspects of the Theory, State of the Art, and Flight Experience of Gravity Gradient Stabilized Vehicles," The Aerospace Corporation, TR-0200(4411)-2, March 1969.
4. Connell, G. M., and V. Chobotov. "Possible Effects of Boom Flutter on the Attitude Dynamics of the OV 1–10 Satellite," Journal of Spacecraft and Rockets, Vol. 6, pp. 90–92, January 1969.
5. Chobotov, V. A. "GSSPS Taking a New Approach to the Space Solar Power Station," Astronautics & Aeronautics, November 1977.
6. Chobotov, V. A. "Photovoltaic, Gravitationally Stabilized, Solid-State Satellite Solar Power Station," Journal of Energy, Vol. 1, No. 6, November/December 1977.
7. Chobotov, V. A. "Gravitationally Stabilized Satellite Solar Power Station in Orbit," Journal of Spacecraft and Rockets, Vol. 14, No. 4, April 1977.
8. Chobotov, V. A. "Radially Vibrating, Rotating Gravitational Gradient Sensor," Journal of Spacecraft and Rockets, Vol. 5, No. 4, 1968.
9. Chobotov, V. A. "Gravitational Excitation of an Extensible Dumbbell Satellite," Journal of Spacecraft and Rockets, Vol. 4, No. 10, 1967.
10. Chobotov, V. A. "Gravity-Gradient Excitation of a Rotating Cable-Counterweight Space Station in Orbit," Journal of Applied Mechanics, Vol. 30, No. 4, December 1963.
11. Misra, A. K., and V. J. Modi. "A Survey on the Dynamics and Control of Tethered Satellite Systems," NASA/AIAA/PSN International Conference on Tethers in Space, Arlington, VA, 17–19 September 1986.
12. von Tiesenhausen, G. "Tethers in Space—Birth and Growth a New Avenue to Space Utilization," NASA TM-82571, February 1984.
13. Tsiolkovsky, K. E. "Speculations Between Earth and Sky," Moscow, Isd-vo AN-SSSR, 1895, p. 35 (reprinted in 1959).
14. Pearson, J. "The Orbital Tower: A Spacecraft Launcher Using the Earth's Rotational Energy," Acta Astronautica, Vol. 2, Nos. 9/10, pp. 785–799, 1975.
15. Clarke, A. C. "The Space Elevator—Thought Experiment or Key to the Universe?" Invited Lecture, 30th Congress of the International Astronautical Federation, Munich, Federal Republic of Germany, September 1979.
16. Chobotov, V. A. "Synchronous Satellite at Less Than Synchronous Satellite Altitude," Journal of Spacecraft and Rockets, Vol. 13, No. 2, pp. 126–128, 1976.
17. Van der Ha, J. C. "Orbital and Relative Motion of a Tethered Satellite System," 34th Congress of the International Astronautical Federation, Budapest, Hungary, preprint IAF-83–320, 1983.
18. Rupp, C. C. "A Tether Tension Control Law for Tethered Subsatellites Deployed Along Local Vertical," NASA TMX-64963, Marshall Space Flight Center, AL, September 1975.
19. Baker, W. P., J. A. Dunkin, et al. "Tethered Subsatellite Study," NASA TMX-73314, Marshall Space Flight Center, Alabama, March 1976.
20. Graciani, F., S. Sgubini, and A. Agneni. "Disturbance Propagation in Orbiting Tethers," NASA/AIAA/PSN International Conference on Tethers in Space, Arlington, VA, September 1986.
21. Kane, T. R., and D. A. Levinson. "Deployment of a Cable-Supported Payload from an Orbiting Spacecraft," Journal of Spacecraft and Rockets, Vol. 14, No. 7, pp. 409–413, 1977.
22. Bainum, P. M., C. M. Diarra, and V. K. Kumar. "Shuttle-Tethered Subsatellite System Stability with a Flexible Massive Tether," Journal of Guidance Control and Dynamics, Vol. 8, No. 2, pp. 230–234, 1985.
23. Breakwell, J. V., and G. B. Andeen. "Dynamics of a Flexible Passive Space Array," J. Spacecraft and Rockets, Vol. 14, No. 9, pp. 556–561, September 1977.
24. Beletskii, V. V., and E. T. Navikova. "On the Relative Motion of Two Cable-Connected Bodies in Orbit," Cosmic Research, Vol. 7, No. 3, pp. 377–384, 1969.
25. Beletskii, V. V. "On the Relative Motion of Two Cable-Connected Bodies in Orbit-II, Cosmic Research, Vol. 7, No. 6, pp. 827–840, 1969.
26. Singh, R. B. "Three Dimensional Motion of a System of Two Cable-Connected Satellites in Orbit," Acta Astronautica, Vol. 18, No. 5, pp. 301–308, 1973.
27. Bergamaschi, S., S. Cusinato, and A. Sinopoli. "A Continuous Model for Tether Elastic Vibrations in TSS," AIAA 24th Aerospace Sciences Meeting, Reno, NV, Paper No. 86–0087, January 1986.
28. Misra, A. K., and V. J. Modi. "Frequencies of Longitudinal Oscillations of Tethered Satellite Systems," AIAA/AAS Astrodynamics Conference, Williamsburg, VA, Paper No. 86–2274, August 1986.
29. Bekey, I., and P. A. Penzo. "Tether Propulsion," Aerospace America, Vol. 24, No. 7, pp. 40–43, 1986.
30. "Selected Tether Applications in Space," Martin Marietta, Contract No. NAS8-35499, February 1985.
31. Colombo, G., M. D. Grossi, et al. "Orbital Transfer and Release of Tethered Payloads," Final Report, Contract NAS8-33691, Smithsonian Institution, Astrophysical Observatory, Cambridge, MA, March 1983.

Chapter 8
Magnetic Stabilization Methods

Use of Earth's magnetic field for attitude control has received much attention from the beginning of the space age. Systems with permanent magnets, hysteresis dampers, current-carrying coils, and ferromagnetic electromagnets have been used for control of spin magnitude and direction in space. Many applications involve the use of a single magnetic dipole aligned with the spin axis or oriented normal to it. Other applications use magnetic torquers in conjunction with dual-spin, momentum-bias, reaction wheel, and control moment gyro systems for attitude control, momentum dumping, or librational damping. Simplified examples of such systems are presented in this chapter.

The magnetic torque $\vec{T}^{(m)}$ acting on a satellite in orbit is the vector cross-product of the spacecraft magnetic dipole moment $\vec{M}$ and the local magnetic induction (magnetic field) or flux density $\vec{B}$, i.e.,

$$\vec{T}^{(m)} = \vec{M} \times \vec{B} \qquad (8.1)$$

where $\vec{M}$ is in ampere-meter^2 (A-m^2), $\vec{B}$ is in webers per meter^2 (WB/m^2), and $\vec{T}^{(m)}$ is newton-meters (n-m).

8.1 Spin Rate (Momentum) Control

The control of the spacecraft spin rate or its angular momentum $\vec{h}$ can be achieved by orienting the magnetic torque vector along the spacecraft angular momentum vector. Thus, for example, if it is desired to decrease (dump) the angular momentum $\vec{h}$ of the spacecraft, then

$$\vec{T}^{(m)} = -k\vec{h}$$
$$= \vec{M} \times \vec{B} \qquad (8.2)$$

where k is a constant of proportionality. The required magnetic moment vector $\vec{M}$ can be found as follows. Multiplying each side of equation 8.2 by $\vec{B}$ yields

$$\vec{B} \times (\vec{M} \times \vec{B}) = \vec{B} \times (-k\vec{h})$$
$$= B^2\vec{M} - \vec{B}(\vec{M} \cdot \vec{B}) \qquad (8.3)$$

where the last term above may be neglected for maximum effectiveness if $\vec{M} \cdot \vec{B} \equiv 0$ when $\vec{M}$ and $\vec{B}$ are orthogonal. Therefore, the required magnetic dipole moment $\vec{M}$ is found as follows:

$$B^2\vec{M} = \vec{B} \times (-k\vec{h})$$
$$\vec{M} = \frac{\vec{B} \times (-k\vec{h})}{B^2} \qquad (8.4)$$

The control torque acting on the spacecraft is of the form

$$\vec{T}^{(m)} = \vec{M} \times \vec{B}$$
$$= -\frac{k}{B^2}[(\vec{B} \times \vec{h}) \times \vec{B}]$$
$$= -\frac{k}{B^2}[B^2\vec{h} - \vec{B}(\vec{h} \cdot \vec{B})] . \qquad (8.5)$$

A schematic diagram of the control system is illustrated in Figure 8.1(a). An example of a magnetically controlled spinning satellite is shown in Figure 8.1(b).

As can be seen from Figure 8.1, the magnetometer determines the spacecraft body components of the geomagnetic field $\vec{B}$. The control logic (computer) computes the required currents for the 1, 2, 3 axes electromagnets such that the resultant $\vec{M}$ vector is normal or nearly normal to the $\vec{B}$ and $\vec{h}$ vectors simultaneously. This reduces inter-axis coupling and enhances the efficiency of the system. In general

$$\vec{M} = iNA\hat{n} \qquad (8.6)$$

where

$$i = \text{current}$$
$$N = \text{number of coil turns}$$
$$A = \text{coil cross sectional area}$$
$$\hat{n} = \text{unit vector normal to coil area}$$

An example of reaction wheel momentum control is illustrated in Figure 8.2(a) for the case of an Earth-oriented spacecraft with a RW in a circular orbit. A half-orbit magnetic dipole switching scheme is also shown in Figure 8.2(b) which ensures that the pitch axis magnetic torque $\vec{T}_p^{(m)}$ is constant during an orbital revolution of the spacecraft.

8.1.1 Precession Control

The control of the spacecraft attitude (orientation) can be achieved in a similar manner to that described in the case of momentum control, except that the magnetic torque vec-

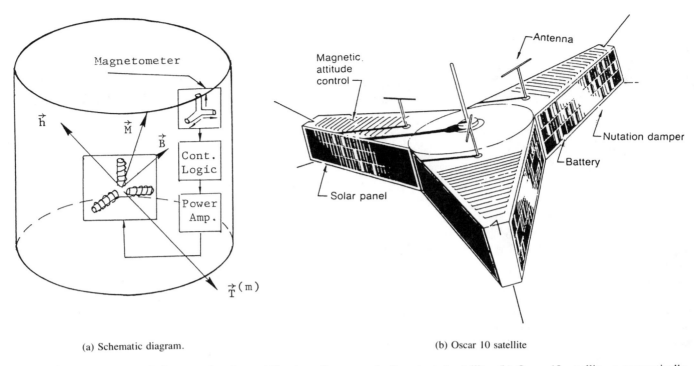

(a) Schematic diagram.   (b) Oscar 10 satellite

Figure 8.1 (a) Functional diagram of active stabilization of a magnetically oriented satellite; (b) Oscar 10 satellite, a magnetically controlled spinning satellite.

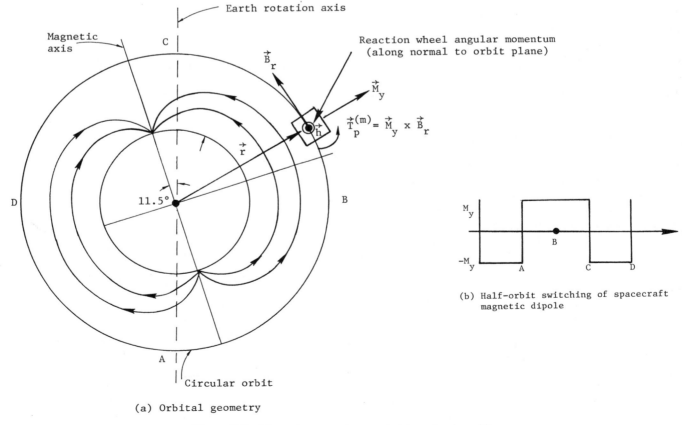

(a) Orbital geometry

(b) Half-orbit switching of spacecraft magnetic dipole

Figure 8.2 Magnetic momentum control in a circular orbit.

tor is ideally directed normal to the spacecraft spin or its angular momentum vector. If, for example, the spacecraft magnetic dipole moment is oriented along the spacecraft spin axis, then a magnetic torque may be generated normal to the spin axis, resulting in a steady precession of the spacecraft angular momentum vector about the Earth's magnetic field vector, as is shown in Figure 8.3.

The spacecraft precession vector $\vec{\Omega}$ can be determined as follows:

Referring to Figure 8.3, the magnetic torque vector $\vec{T}^{(m)}$ is given by the expression

$$\vec{T}^{(m)} = \vec{M} \times \vec{B}$$
$$= \frac{d\vec{h}}{dt} . \qquad (8.7)$$

Orienting the magnetic dipole vector $\vec{M}$ along the spacecraft angular momentum vector $\vec{h}$, allows us to write

$$\vec{M} = M \frac{\vec{h}}{h} \qquad (8.8)$$

and equation 8.7 becomes

$$\frac{d\vec{h}}{dt} = \vec{\Omega} \times \vec{h}$$
$$= \vec{M} \times \vec{B}$$
$$= M \frac{\vec{h}}{h} \times \vec{B}$$
$$= -\frac{M\vec{B}}{h} \times \vec{h} \qquad (8.9)$$

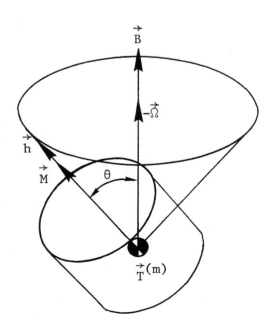

Figure 8.3 Magnetic precession of the spacecraft's angular momentum vector about the Earth's magnetic field $\vec{B}$.

which shows that the precession is

$$\vec{\Omega} = -\frac{M\vec{B}}{h} . \qquad (8.10)$$

This result is valid for rapidly spinning satellites when the angular momentum vector $\vec{h}$ is generally aligned along the spin axis. For rapidly spinning satellites (i.e., those for which the orbital period is very much greater than the spin period), only the spin axis component of the satellite's dipole moment will produce a torque that does not have a zero average value over a spin period. This torque (which is orthogonal to both the spin axis and the magnetic field vector) tends to cause precession of the spin axis.

When the satellite attitude, with respect to an inertial reference frame, varies infinitesimally throughout an orbit (i.e., a spin stabilized or a stellar or solar oriented satellite), the average torque over an orbit is often important. Because $\vec{M}$ is fixed, the problem becomes that of determining an average value for $\vec{B}$. For this computation, the tilted dipole model for the Earth's magnetic field can be employed.

8.1.2 Sun Synchronous Spacecraft Example

A dual-spin spacecraft in a Sun synchronous orbit is required to rotate (precess) approximately one degree per day (360° per year) in order to maintain solar cell orientation normal to the Sun. Assuming a circular 600 kilometer altitude orbit with a 98-degree inclination, it is desired to determine the magnetic dipole moment strength M oriented along the spin axis for a satellite having a spin axis inertia $I = 2.7 \times 10^9$ gm · cm^2 and a spin rate $\omega_s = 1.15$ rad/s.

The average north-south geomagnetic field strength for this orbit is $B_{av} \approx 0.115$ gauss. The required dipole moment, as a function of angular momentum h, is $M = \Omega h / B_{av}$ where $\Omega \approx 1$ degree per day (2×10^{-7} rad/s) is the precession rate orthogonal to the spin axis of the satellite. Thus,

$$h = I\omega_s = 3.11 \times 10^9 \frac{\text{gm} \cdot \text{cm}^2}{\text{sec}}$$

and

$$M = \frac{\Omega h}{B_{av}}$$
$$= 5463 \text{ pole} \cdot \text{cm} .$$

8.1.3 Passive Spin Rate Control

When the orientation of the spacecraft or a portion of the spacecraft moves with respect to the magnetic field vector, torques caused by induced currents (eddy currents) and the irreversible magnetization of permeable materials (hysteresis effects) must be considered. In general, an accurate assessment of these torques is extremely difficult and simplifying assumptions regarding the shape of the spinning section, its magnetic characteristics, and the nature of the interaction with the ambient field are required to approximate the magnitude of these effects.

In most practical situations, the rotating or moving sections in which eddy currents flow will consist of a structural material that has a permeability very nearly equal to that of free space. While eddy currents are also generated in a rotating permeable material, the resulting torques are usually negligible compared to the torques resulting from magnetization of the material. The following equations for the estimation of eddy current torques on simple geometric figures are based on the assumption that the permeability of the material is equal to that of free space and that the magnetic field caused by the induced currents is so small compared to the ambient field that the field remains uniform. For objects rotating with very high angular velocity (reaction wheels, for example), the latter assumption may be invalid and the equations below will overestimate the despin torque [1].

8.1.3.1 Eddy Current Effects

Total (external) torque $\quad \vec{T}_e = k_e(\vec{\omega} \times \vec{B}) \times \vec{B} \qquad$ N · m

Despin component of torque $\quad T_{se} = -k_e(B_\perp)^2 \omega_s \qquad$ N · m

Precession component of torque $\quad T_{\perp e} = k_e \omega_s B_s B_\perp \qquad$ N · m

Spin acceleration caused by eddy currents $\quad \dfrac{\Delta \omega_s}{\Delta t} = -\dfrac{k_e}{I_s}(B_\perp)^2 \omega_s \qquad$ rad/s^2

where

$\dfrac{\Delta \omega_s}{\Delta t} =$ change in spin rate in radians per second^2

$\vec{B} =$ ambient magnetic flux density in teslas or webers per meter^2

$B_\perp =$ component of $\vec{B}$ orthogonal to spin axis in tesla

$B_s =$ component of $\vec{B}$ parallel to spin axis in tesla

$\vec{\omega} =$ spacecraft angular velocity vector

$\omega_s =$ spin angular velocity in radians per second

$I_s =$ spacecraft spin axis moment of inertia in kilogram · meters^2

$k_e =$ a constant which depends on the geometry and electrical conductivity of the rotating object

For a thin spherical shell of radius r, thickness d, and conductivity σ

$$k_e = \frac{2\pi}{3} r^4 \sigma d$$

For a circular loop, located in a plane through the spin axis, with radius r, cross-sectional area S, and conductivity σ

$$k_e = \frac{\pi}{4} \sigma r^3 S$$

For a thin walled cylinder, with length L, radius r, thickness τ, and conductivity σ

$$k_e = \pi \sigma r^3 L\tau \left(1 - \frac{2\tau}{L} \tanh \frac{L}{2\tau}\right).$$

For a thin walled cylinder spinning about a transverse axis through the geometric center, the average torque is one-half of that produced by rotation around the cylinder axis.

8.1.3.2 Hysteresis Damping

When a magnetically permeable material is rotated in a magnetic field, energy is dissipated because of the motion of the magnetic domains. The energy loss in joules per cycle over any complete period of rotation is constant and is given by the following integral over the area of the hysteresis loop

$$\Delta E_h = V \oint H \, dB_i$$

where

$V =$ volume of the material in cubic meters

$H =$ the intensity of the external field in amperes per meter

$B_i =$ the induced magnetic flux in the magnetic material

$\oint H \, dB_i =$ area of hysteresis loop.

Because the energy loss per cycle is independent of spin rate, the spin rate will decrease linearly with time and can actually go to zero (with respect to the ambient field) in a finite length of time.

8.2 Control and Minimization of Magnetic Disturbance

When it is necessary to ensure that the magnetic disturbance torques do not exceed a specified value, procedures must be established to control the spacecraft dipole moment and other magnetic torque sources, such as current loops and permeable materials.

When a maximum value for the spacecraft dipole moment is specified, it is essential that a control procedure be initiated at the start of the program and continued throughout design and development. Experience with past spacecraft development programs has shown that the institution of controls midway in the program (to "clean up" a magnetically "dirty" spacecraft) is costly and generally ineffective. In particular, basic precautions for the avoidance of current loops and unnecessary magnetic materials must be incorporated in the design phase.

Examples A and B below illustrate the estimation of the magnetic dipole moment $\vec{M}$ and the resultant torque $\vec{T}^{(m)}$ on a satellite in a geosynchronous equatorial orbit when current loops are present in the solar arrays.

Example A. Magnetic Dipole

A typical spacecraft solar array can generate 1800 watts at 30 volts using six identical solar panels. A current of 8.4 amperes per panel is thus obtained. Each panel consists of 18 strings of solar cells, 15 of which are 2 meters long and are connected in parallel.

A single return wire is used which results in a current loop of approximately 1.11 square meters. A current of 8.4 amps flows in the return wire; the magnetic moment M due to this current loop is given by the equation

$$M = NiA$$

where

$$N = \text{number of wire turns;}$$
$$i = \text{current} = 8.4 \text{ amps}$$
$$A = \text{current loop area} = 1.11 \text{ m}^2.$$

Thus $M = NiA = 9.35$ amp-turn $\cdot$ m^2 = 9350 pole $\cdot$ cm for a single turn ($N = 1$).

Example B. Magnetic Disturbance Torque

The magnetic disturbance torque $\vec{T}^{(m)}$ on the satellite resulting from the interaction between a magnetic dipole moment $\vec{M}$ and the Earth's magnetic field vector $\vec{B}$ is given by the equation

$$\vec{T}^{(m)} = \vec{M} \times \vec{B} \, .$$

Since $\vec{M}$ is directed normal to the solar arrays, the maximum value of the disturbance torque occurs when $\vec{B}$ is parallel to the solar arrays. At the synchronous equatorial orbit $B \approx 10^{-3}$ gauss $= 10^{-7}$ weber/m^2 which is equivalent to 100 gammas on an average basis using a simple dipole model of the Earth's magnetic field. The maximum disturbance torque acting on the satellite due to six identical solar panels is therefore

$$T^{(m)}_{\max} = MB$$
$$= 6 \times 9.35 \times 10^{-7} \text{ newton} \cdot \text{meters} \, .$$

This torque can be eliminated by twisting the solar array strings (wires) to reduce the current loop area to zero.

8.2.1 Materials

Close control over the construction materials is an important consideration and must include: (1) knowledge of the materials, (2) knowledge of the suppliers, (3) documentation of approved materials and sources, and (4) inspection. In addition, preference should be given to magnetically "hard" over magnetically "soft" materials, and long and narrow shapes are to be avoided. A "hard" magnetic material means a material in which the magnetic moment is essentially unchanged by small changes in the field around it, whereas, conversely, the magnetic state of a "soft" material is predominantly determined by the ambient field. Generally, "hard" magnetic materials are also physically hard (i.e., a high Brinell hardness number) and conversely for magnetically soft materials. Alnico V and cobalt steel are magnetically hard, while mu-metal (for high permeability μ) and mild steel are examples of soft magnetic materials.

Stainless steel provides a particularly difficult set of problems because even "nonmagnetic" stainless steel can become magnetic when machined. This is a processing effect.

8.2.2 Components

In some functional spacecraft components, use of magnetic materials is unavoidable. Typical items that generate substantial dipole moments are current driven devices such as traveling wave tubes, tape recorders, latching relays, batteries, motors, gyros, coaxial switches, photomultiplier tubes, and solenoid valves.

Components having large dipole moments should be paired to produce magnetic moment cancellation, or if this is not feasible, they should be individually compensated. It must be recognized that because of temporal instabilities of the dipole moment of equipment or the compensating magnets, complete cancellation is not practical.

Shielding, i.e., enclosing the offending component in a container of highly permeable material (such as mu-metal), can create problems in the functioning of the shielded component. Shielding should, therefore, be avoided as a means of reducing the dipole moment of a component containing functional magnets. If essential, a shield can be used to reduce the external effects of dipole moment variations caused by moving magnetic parts. A preferred practice, if functionally feasible, is to mount a compensating magnet on the moving part.

Small components used in large quantities often lead to significant dipole moments. Thus, many electronic components, such as transistors and capacitors, have nickel cased and nickel plated leads; connectors contain mostly nonmagnetic material, but occasionally magnetic lock washers and compression washers find their way into the assembly; also shielding on coaxial cables will often be magnetic. Wire should, in general, be checked because nickel plating of conductors is a common practice.

8.2.3 Current Loops

Current flowing in the solar array can be a major contributor to the spacecraft dipole moment as has been shown in Section 8.2 unless proper wiring is included in the design. The recommended technique is to route current return wires directly behind the solar cells (backwiring). This minimizes loop area, and the technique is effective when portions of the array will be switched on and off. Twisting of leads in wiring harnesses to cancell induced magnetic fields is also recommended. However, there is a compromise between reliability and minimization of dipole moment that must be considered, because twisting of the wires may lead to insulation failure.

Ground currents are often neglected as a potential source of current loops. Unless care is exercised in grounding, particularly through the use of single point grounding, a current loop of substantial area may exist.

Batteries, regardless of the material used for the case or the electrodes, will have a dipole moment caused by current flow. To avoid this, batteries should be used in pairs so as to produce dipole moment cancellation or, if an odd number is essential, a loop of wire in series with the battery can be used to cancel the resulting magnetic moment.

8.2.4 Dipole Determination

Tests that can be performed on a spacecraft depend to a large extent on the available facilities. When possible, a measurement of the dipole moment should be obtained in the deployed or orbital configuration with all systems activated.

Generally, the spacecraft dipole moment is determined using various test methods. Measurements of magnetic properties are based either on torque measurement or magnetic field plots. For the assessment of disturbance torques, methods based on torque measurement are preferable.

Compensation consists of affixing to the spacecraft a permanent magnet whose dipole moment is equal and opposite to that of the spacecraft. The use of three orthogonal components of the dipole moment is an equivalent procedure. After the compensation magnets are installed, the measurement should be repeated. This is due to the fact that the addition of the permanent magnets will cause a change in induced effects. Thus, the dipole moment, after addition of the compensating magnets, will rarely correspond to the difference between the initially measured moment and the compensating moment. Further, the value achieved will normally depend on the location of the compensating magnets. When the initial dipole moment is 1 A · m² (1000 pole · cm) or greater, compensation is best accomplished in several stages, using small trim magnets for the final stage.

Depending on the phase of the spacecraft development program, any one of the following situations may exist:

1. The spacecraft dipole moment is known from measurements made on the flight hardware.
2. The spacecraft dipole moment is estimated on the basis of measurements made on similar equipment.
3. Little similarity with previous spacecraft exists and the estimate of dipole moment must be based on other known parameters of the spacecraft such as mass, mission requirements, and on-board equipment.

These situations are listed in the order of increasing uncertainty regarding the value of $\vec{M}$. Generally, the dipole moment will be measured at some point in the program. For situations 2 and 3, only when analysis has shown the magnetic disturbance torque to have negligible effect on control system performance, will it be possible to consider the elimination of direct measurements.

When situation 3 occurs and the designer must establish an initial estimate for the spacecraft dipole moment, it is recommended that the estimate be based on the factors given in Table 8.1. The factors are, of course, related to the stringency of the magnetic properties control program.

For a spinning spacecraft, the estimated value of M obtained from Table 8.1 is the component along the spin axis. For nonspinning spacecraft, the dipole orientation must either be determined from available information regarding equipment location or the worst case orientation should be assumed, i.e., the direction that imposes the maximum torque burden on the system.

The numerical factors shown in Table 8.1 for Class III require that sources of unusually large dipole moments be avoided or individually compensated and that the use of materials subject to induced magnetism be minimized. Without these precautions, the dipole moment per unit mass may be a factor of 10 or more greater than the nominal value shown in the table.

The estimate of the spacecraft dipole moment obtained from Table 8.1 may be improved through a comparison of the planned spacecraft with past or current spacecraft for which magnetic properties have been established. The criteria for magnetic properties control are given in Table 8.2.

8.3 Recommended Practice for the Control of the Spacecraft Magnetic Moment

The following general recommendations are suggested to enhance magnetic cleanliness of the spacecraft. The recommended practice applies to the selection of the spacecraft materials, components, current loops, and the dipole determination methods.

Table 8.1 Factors for Estimating Spacecraft Dipole Moment (M) [1]

Category of Magnetic Properties Control	Estimate of Dipole Moment per Unit Mass for (a) Nonspinning Spacecraft (b) Spinning Spacecraft	
	(a) A · m²/kg	(b) A · m²/kg
Class I	1×10^{-3}	0.4×10^{-3}
Class II	3.5×10^{-3}	1.4×10^{-3}
Class III	10×10^{-3} and higher	4×10^{-3} and higher

Table 8.2   Criteria for Magnetic Properties Control [1]

	Class I	Class II	Class III
Design	Formal specification on magnetic properties control; approved materials and parts lists; cancellation of moments by preferred mounting arrangements and control of current loops.	Advisory specifications and guidelines for material and parts selection. Avoidance of "soft" magnetic materials or current loops and awareness of good design practices.	Nominal control over current loops; guidelines for avoidance of "soft" magnetic materials.
Quality control	Complete magnetic inspection of parts and testing of subassemblies.	Inspection or test of suspect parts.	Test of subassemblies that are potentially major sources of magnetic dipole moment.
Test and compensation	"Deperming" either at subassembly or spacecraft level; test of final spacecraft assembly and compensation if required.	"Deperming" and compensation frequently used.	Test and compensation optional.

NOTE:
Class   I—Magnetic torques dominant when compared with other torques.
       II—Magnetic torques comparable to other torques.
      III—Magnetic torques insignificant when compared with other torques.

Materials:

1. Use magnetically hard materials.
2. Know your supplier and inspect supplied materials carefully.

Components:

1. Limit the number of magnetic field producing devices such as traveling wave tubes, batteries, motors, tape recorders, switches, gyros, latching relays, solenoid valves, photomultiplier tubes, etc.
2. Pair components with large dipole moments or cancell their magnetic moments with small permanent magnets.
3. Avoid small components in large quantities such as transistors, nickel plated leads, shielding on coaxial cables, magnetic lock or compression washers, etc.

Current Loops:

1. Backwire solar arrays to reduce current loop areas.
2. Twist leads in harnesses.
3. Use single point grounding to reduce area of current loops.
4. Pair batteries to help cancel current flow dipole moments.

Dipole Determination:

1. Activate all systems in a testing facility to simulate operating conditions.
2. Use measurements based on torques or magnetic fields.
3. Compensate the spacecraft's magnetic dipole moment with the addition of permanent magnets.
4. Estimate the dipole moment using magnetic moment data of similar equipment.

8.4   References

1. "Spacecraft Magnetic Torques," NASA Space Vehicle Design Criteria, NASA SP-8010, March 1969.
2. Wertz, J. R., ed. *Spacecraft Attitude Determination and Control*, D. Reider Publishing Co., 1980.
3. Schmidt, G. E., Jr. "The Application of Magnetic Attitude Control to a Momentum Biased Synchronous Communications Satellite," AIAA Paper No. 75–1055, 20 August 1975.
4. Takezawa, S., and K. Ninomiya. "A New Approach to the Analysis and Design of Magnetic Stabilization of Satellites," Acta Astronautica, Vol. 7, pp. 731–751, 1980.
5. Tossman, B. E., et al. "MAGSAT Attitude Control System Design and Performance," AIAA Paper No. 80–1730, 1980.
6. Tossman, B. E. "Magnetic Attitude Control System for the Radio Astronomy Explorer—A Satellite," AIAA Paper No. 68–855.
7. Spencer, T. M. "Automatic Magnetic Control of a Momentum-Biased Observatory in Equatorial Orbit," Journal of Spacecraft and Rockets, Vol. 14, No. 4, April 1977, pp. 193–194.
8. Mobley, F. F., Konigsberg, R., and G. H. Fountain. "Attitude Control System of the SAS-C Satellite," AIAA Paper No. 74–901.
9. Foulke, H. F., Siegel, S. H., and A. Das. "Stability Criterion for Passive Magnetically Anchored Rate Dampers," Journal of Spacecraft and Rockets, Vol. 14, No. 4, April 1977, pp. 195–196.

Chapter 9

Stability of Motion

The theory of the stability of motion relates to the basic concepts of dynamics and mathematics in general. Comprehensive treatments of it can, for example, be found in references 1 through 8 as well as in many other sources. The concepts of stability, as related to spacecraft attitude control, deal with the stability of equilibrium (in the absence of closed loop control systems) and with transient or steady-state motion. It is the purpose of this chapter to examine the concepts of stability as they apply to spacecraft attitude dynamics and control and to present some analytical techniques for their analysis.

9.1 Types of Stability Concepts

The word *stable* (derived from the Latin adjective *stabilem*, to stand firmly) has many shades of meaning and therefore must be precisely defined if it is to be of use in the study of spacecraft attitude dynamics and control. Two such stability concepts are related to linear and nonlinear systems. The first concept is that of "infinitessimal" or Lagrange stability, which is a "boundedness" concept applicable to linear systems. If, for example, a small deviation from some equilibrium point remains bounded, then the motion is said to be Lagrange, or infinitesimally stable. This is, in general, applicable to systems which can be described by linear differential equations, the equilibria or the solutions of which are either stable or asymptotically stable when the perturbed motion approaches the equilibrium condition at some time in the future.

The second stability concept requires that a solution to a differential equation, beginning sufficiently close to the equilibrium (origin), must remain arbitrarily close to itself after a perturbation. This concept is that due to A. Liapunov and is applicable to the solutions described by both linear and nonlinear differential equations.

Lagrange stability thus required only that the solution (trajectory) remain within a finite distance from equilibrium while Liapunov stability requires arbitrarily small deviations after a perturbation from equilibrium. Therefore, Liapunov stability implies Lagrange stability, but the opposite is not true. An illustration of Lagrange and Liapunov stability concepts is provided by the pendulum phase plane diagram (shown in Figure 9.1). The bottom rest position of the pendulum, for example, responds to a small displacement or velocity perturbation with a small oscillation about bottom rest, which continues indefinitely in the absence of damping. The oscillation can be made as small as desired by keeping the perturbation sufficiently small. The position of bottom rest (equilibrium) is therefore Liapunov and also Lagrange stable since it is bounded. The addition of damping results in asymptotic stability of the solution.

The position of top rest is Liapunov unstable because a non-zero deviation in attitude (θ) or attitude rate ($\dot{\theta}$) produces a dramatic change in θ no matter how small the initial perturbation. Since θ is unbounded for any initial value of $\dot{\theta}$ (from the position of top rest), the latter is also Lagrange unstable.

The oscillatory motion about bottom rest $\theta = 0, 2\pi, 4\pi$, etc. is Liapunov unstable because the period depends on amplitude and if a small perturbation in amplitude results, then a grossly different position of the perturbed pendulum will result compared to the unperturbed oscillatory position. The difference between the perturbed and unperturbed amplitude cannot vary by more than a factor of two; however, it is bounded and the unperturbed motion is therefore Lagrange stable, as can be seen in Figure 9.1

9.2 Stability Criteria for Linear Systems

9.2.1 Routh-Hurwitz Conditions

A linear system is said to be stable if and only if the roots of its characteristic equation have negative real parts. If any root has a real part that is positive, the system is asymptotically unstable. For an asymptotically stable system, the state variables remain in the neighborhood of equilibrium and eventually return to the equilibrium state when the system is given an initial small perturbation. If any root has a real part that is zero, the system is again unstable (or conditionally stable).

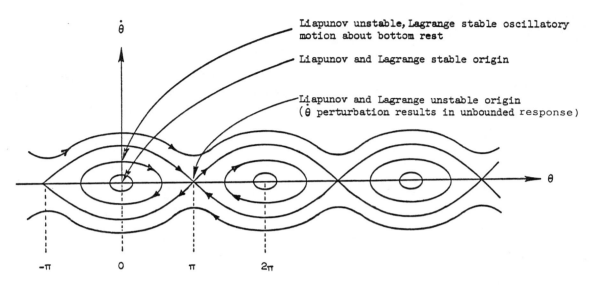

Figure 9.1 Pendulum phase plane diagram.

The Routh-Hurwitz theorems state that the necessary condition for the roots of the characteristic equation [7, 9, 10]

$$a_o s^n + a_1 s^{n-1} + \ldots a_{n-1} s + a_n = 0 \qquad (9.1)$$

to have negative real parts is that all coefficients must be nonzero and of the same sign, i.e., $a_i > 0$, $i = 0, \ldots$, n. Additional necessary and sufficient conditions, as given in reference 10, are

$$\Delta_1 = a_1 > 0, \qquad (9.2)$$

$$\Delta_2 = \begin{vmatrix} a_1 & a_o \\ a_3 & a_2 \end{vmatrix} > 0 \qquad \Delta_3 = \begin{vmatrix} a_1 & a_o & 0 \\ a_3 & a_2 & a_1 \\ a_5 & a_4 & a_3 \end{vmatrix} > 0$$

$$\Delta_4 = \begin{vmatrix} a_1 & a_o & 0 & 0 \\ a_3 & a_2 & a_1 & a_o \\ a_5 & a_4 & a_3 & a_2 \\ a_7 & a_6 & a_5 & a_4 \end{vmatrix} > 0$$

$$\qquad (9.3)$$

$$\Delta_5 = \begin{vmatrix} a_1 & a_o & 0 & 0 & 0 \\ a_3 & a_2 & a_1 & a_o & 0 \\ a_5 & a_4 & a_3 & a_2 & a_1 \\ a_7 & a_6 & a_5 & a_4 & a_3 \\ a_9 & a_8 & a_7 & a_6 & a_5 \end{vmatrix}$$

$$\Delta_i = \begin{vmatrix} a_1 & a_o & 0 & 0 & \ldots \\ a_3 & a_2 & a_1 & a_o & \ldots \\ a_5 & a_4 & a_3 & a_2 & \ldots \\ \ldots & \ldots & \ldots & \ldots & \ldots \\ a_{2i-1} & a_{2i-2} & \ldots & \ldots & a_i \end{vmatrix} > 0 . (9.4)$$

9.2.2 Alternate Conditions for Second Order Systems

An alternative criterion for the equilibrium of second order differential equations of the form

$$\ddot{x} = f(x) \qquad (9.5)$$

results when $f(x_o) = 0$ where x_o is the equilibrium condition. A small variation from equilibrium $x = x_o + \xi$ results in

$$\ddot{x} = \ddot{\xi} = f(x_o + \xi)$$

$$= f(x_o) + \xi f'(x_o) + \frac{1}{2} \xi^2 f''(x_o) + \ldots . \qquad (9.6)$$

by Taylor's expansion about equilibrium where dots denote differentiation with respect to time and primes with respect to x. For example,

$$f'(x_o) = \left. \frac{df(x)}{dx} \right|_{x = x_o} . \qquad (9.7)$$

For small ξ the variational (perturbation) equation becomes

$$\ddot{\xi} - f'(x_o)\xi = 0 \qquad (9.8)$$

where the stability of ξ (and therefore of equilibrium) is assured when $f'(x_o) < 0$ and instability when $f'(x_o) > 0$. For example, consider pendulum equation of the form

$$\ddot{\theta} + \frac{g}{\ell} \sin \theta = 0 \qquad (9.9)$$

where θ is the deviation from bottom rest, ℓ is the pendulum length, and g is the gravitational acceleration. Therefore,

$$\ddot{\theta} = -\frac{g}{\ell} \sin \theta$$

$$= f(\theta)$$

and

$$f'(\theta) = -\frac{g}{\ell} \cos \theta . \qquad (9.10)$$

For the position of bottom rest $(\theta = 0), f'(0) = -g/\ell < 0$; therefore, it is Lagrange stable. For the position of top rest $(\theta = \pi), f'(\pi) = g/\ell > 0$ and thus it is both Lagrange and Liapunov unstable.

9.2.3 Quasilinear Systems

Systems described by equations with periodic coefficients (quasilinear systems) may be described by a matrix equation of the type

$$\frac{dx}{dt}(\tau) = A(\tau) x(\tau) \tag{9.11}$$

where $x(\tau)$ is a $n \times 1$ column matrix (or vector) and $A(\tau)$ is an $n \times n$ matrix of known periodic coefficients with period T. Here τ is a time dependent variable (e.g., $\tau = \omega t$). A numerical procedure (Floquet theory) can be used to determine the stability of the zero (trivial) solution of equation 9.11 for the special case $A(\tau) = A(\tau + T)$ when each element of $A(\tau)$ is either periodic (with period T) or constant.

Second order differential equations of the type

$$\frac{d^2x}{d\tau^2} + q(\tau) \frac{dx}{d\tau} + r(\tau)x = 0 \tag{9.12}$$

where $q(\tau + T) = q(\tau)$ and $r(\tau + T) = r(\tau)$ can be transformed using

$$x(\tau) = y(\tau) \exp\left[-\frac{1}{2} \int_o^\tau q(\alpha)d\alpha \right] \tag{9.13}$$

to yield Hill's equation

$$\frac{d^2y}{d\tau^2} + p(\tau)y = 0 \tag{9.14}$$

where

$$p(\tau) = [r(\tau) - \frac{1}{2}\frac{dq(\tau)}{d\tau} - \frac{1}{4}q^2(\tau)] = p(\tau + T) .$$

A special case of Hill's equation is the Mathieu equation

$$\frac{d^2y}{d\tau^2} + (a + 16q \cos 2\tau)y = 0 \tag{9.15}$$

where a and q are real. Solutions of equation 9.15 are known as Mathieu functions (obtained by E. Mathieu in 1868 prior to the development of the Floquet theory in 1883).

For example, consider the following particular solution of equation 9.15

$$y = e^{\mu\tau}\phi(\tau) \tag{9.16}$$

where $\phi(\tau)$ is periodic in τ with period π or 2π. Since this solution is unchanged if $-\tau$ is written for τ, then $y = e^{-\mu\tau}\phi(-\tau)$ is another independent solution. Therefore, the general solution is

$$y = c_1e^{\mu t}\phi(\tau) + c_2e^{-\mu t}\phi(\tau) \tag{9.17}$$

where c_1 and c_2 are arbitrary constants [5]. The solution is stable if μ is imaginary and unstable if μ is real.

The stability of solutions is shown in Figure 9.2 in the $a - q$ plane. It can be seen from this figure that instability occurs at specific values of a (e.g., 1, 4, 9, etc.) for any nonzero values of the parameter q. The parametric instability regions are somewhat reduced if damping is present in the system.

9.3 Stability in Nonlinear Systems

Linear relationships are usually approximations of nonlinear differential equations. The latter provide a better description of physical phenomena (e.g., Euler's equations) but are often not solvable analytically with only occasional exceptions. A general approach for the investigation of the stability of equilibrium points in nonlinear systems is provided by the "Direct" or "Second Method of Liapunov." In this approach the solutions of the nonlinear differential equations are not required to determine Liapunov stability or instability of an equilibrium state. The method is based on the construction of a quadratic form (a Liapunov function V) the properties of which determine the stability or instability of the solutions to the differential equations of motion. The size of the stability region can also be determined since the dynamical response of the nonlinear system is generally dependent on the initial conditions.

The principal idea of the "Second Method" is expressed in the following physical reasoning: If the rate of change $dE(x)/dt$ of the energy $E(x)$ of an isolated physical system is negative for every possible state x, except for a single equilibrium state x_e, then the energy will continually decrease until it finally assumes its minimal value $E(x_e)$. For example, a dissipative system perturbed from equilibrium will return to its original state. This is expressed mathematically by stating that a system is stable if and only if there exists a Liapunov function $V(x) > 0$ whose time derivative $\dot{V}(x) < 0$ when $x \neq x_e$, and when $x = x_e$,

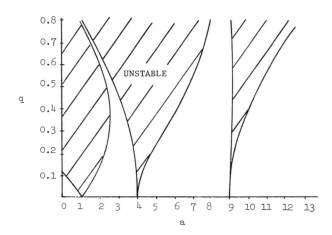

Figure 9.2 Stability chart for Mathieu's equation.

$V(x) = \dot{V}(x) = 0$. Often, but not always, $V = E$, where E is the energy of the system [8].

The concept of stability and the theorems of the direct or second method of Liapunov are presented in the sections that follow. Additional information on this topic can be found in references 1–12.

9.3.1 Matrix Differential Equations

The system of differential equations

$$\frac{dx_i}{dt} = \dot{x}_i = f_i(x_1, x_2, \ldots, x_n, t)(i = 1, 2, \ldots, n) \tag{9.18}$$

is easily expressed as a first order matrix differential equation, namely,

$$\dot{X} = f(X, t) \tag{9.19}$$

where

$$X = \begin{pmatrix} x_1 \\ x_2 \\ \cdot \\ \cdot \\ \cdot \\ x_n \end{pmatrix}$$

and

$$f(X, t) = \begin{pmatrix} f_1(x_1, \ldots, x_n, t) \\ \cdot \\ \cdot \\ \cdot \\ f_n(x_1, \ldots, x_n, t) \end{pmatrix}$$

This is a general (nonlinear, nonautonomous) first order matrix differential equation. If

$$f(X, t) = f(X) \tag{9.20}$$

then the equation is autonomous; if f is a linear function of X, it is a linear autonomous matrix differential equation. Thus, a single n-th order differential equation

$$\frac{d^n x}{dt^n} = f\left(x, \ldots, \frac{d^{n-1}x}{dt^{n-1}}, t\right) \tag{9.21}$$

assumes the form

$$\dot{X} = AX + bf(X, t) \tag{9.22}$$

where

$$X = \begin{pmatrix} x_1 \\ x_2 \\ \cdot \\ \cdot \\ \cdot \\ x_n \end{pmatrix}, \quad A = \begin{pmatrix} 0 & 1 & 0 & \ldots & 0 \\ 0 & 0 & 1 & 0 & 0 \\ \cdot & & & & \cdot \\ \cdot & & & & \cdot \\ \cdot & & & & \cdot \\ 0 & & & & 0 \end{pmatrix} \tag{9.24}$$

and

$$b = \begin{pmatrix} 0 \\ \cdot \\ \cdot \\ \cdot \\ 0 \\ 1 \end{pmatrix}, \quad f(X, t) = f(x_1, x_2, \ldots x_{n-1}, t) . \tag{9.24}$$

When

$$x_1 = x, \ x_2 = \frac{dx}{dt}, \ \ldots x_n = \frac{d^{n-1}x}{dt^{n-1}} \tag{9.25}$$

and

$$\dot{x}_1 = x_2, \dot{x}_2 = x_3, \ldots \dot{x}_n = f(x_1, x_2, \ldots, x_n t) , \tag{9.26}$$

then one is dealing with a first-order, nonlinear, nonautonomous matrix differential equation with a linear part.

EXAMPLE: Consider,

$$\ddot{y} + a\dot{y} + by = 0 . \tag{9.27}$$

Let,

$$\begin{aligned} y &= x_1 \\ \dot{y} &= x_2 \end{aligned} \tag{9.28}$$

then,

$$\begin{aligned} \dot{x}_1 &= x_2 \\ \dot{x}_2 &= -ax_2 - bx_1 \end{aligned} \tag{9.29}$$

or, in matrix form

$$\begin{pmatrix} \dot{x}_1 \\ \dot{x}_2 \end{pmatrix} = \begin{pmatrix} 0 & 1 \\ -b & -a \end{pmatrix} \begin{pmatrix} x_1 \\ x_2 \end{pmatrix} . \tag{9.30}$$

This is of the form

$$\dot{X} = AX \tag{9.31}$$

where the matrix A can be expressed as

$$A = \begin{pmatrix} 0 & 1 \\ -b & -a \end{pmatrix} \tag{9.32}$$

and

$$X = \begin{pmatrix} x_1 \\ x_2 \end{pmatrix} . \tag{9.33}$$

9.3.2 Some Definitions

Consider an arbitrary dynamic system and assume its perturbed motion may be described by a system of differential equations which may be reduced to the normal form.

In the following, certain real functions $V(x)$ are considered which are defined within a certain vicinity of the origin of the coordinates.

9.3.2.1 Definition 1

The function $V(x)$ is definite (positive or negative) within the region H if for

$$|x| \leqslant H$$

where H is a sufficiently small positive number $V(x)$ assumes values of fixed sign or zero only when $x = 0$.

EXAMPLE: Consider two functions $V_1(x)$ and $V_2(x)$ where

$$V_1(x) = x_1^2 + x_2^2 + x_3^4$$
$$V_2(x) = x_1^2 + 2x_1x_2 + 2x_2^2 + x_3^2$$
$$= (x_1 + x_2)^2 + x_2^2 + x_3^2 \qquad (9.34)$$

Both functions are positive definite.

Now consider

$$V(x) = x_1^2 + x_2^2 + x_3^2(1 - x_3) \qquad (9.35)$$

Here $V(x)$ is positive definite only for $|x_3| < 1$ or when H is sufficiently small.

9.3.2.2 Definition 2

The function $V(x)$ is semi-definite (positive or negative) if, in the region H where $|x| \leqslant H$, it assumes a value of fixed sign or is zero [$V(x)$ also may become zero for $x \neq 0$].

EXAMPLE: Consider

$$V(x) = x_1^2 + x_2^2 - x_3^4 \qquad (9.36)$$

Here $V(x)$ is a semi-definite function.

9.3.2.3 Criteria for Definiteness of Functions

The general case is very complex, but particular cases can be investigated by the following criteria. Let V be a quadratic function of x (second-order homogeneous polynomial) as follows:

$$2V = \sum_{\alpha, \beta = 1}^{n} C_{\alpha\beta} x_\alpha x_\beta$$
$$= C_{11}x_1^2 + C_{22}x_2^2 + \ldots + C_{12}x_1x_2$$
$$+ C_{13}x_1x_3 + \ldots$$

$$= [x_1 x_2 \ldots x_n] \begin{pmatrix} C_{11} & \ldots & C_{1n} \\ \cdot & & \\ \cdot & & \\ \cdot & & \\ C_{n1} & \ldots & C_{nn} \end{pmatrix} \begin{pmatrix} x_1 \\ x_2 \\ \cdot \\ \cdot \\ \cdot \\ x_n \end{pmatrix}$$

$$= x'Px \qquad (9.37)$$

where

$$P = \begin{pmatrix} C_{11} & \ldots & C_{1n} \\ \cdot & & \\ \cdot & & \\ \cdot & & \\ C_{n1} & \ldots & C_{nn} \end{pmatrix}. \qquad (9.38)$$

Since $x_ix_j = x_jy_i$, $a_{ij} = a_{ji}$, and P is a symmetric matrix.

9.3.2.4 Silvester Condition

In order for the quadratic form (equation 9.37) to be positive definite it is necessary and sufficient that the principal minors of its discriminant be positive. That is,

$$c_{11} > 0, \quad \begin{vmatrix} c_{11} & c_{12} \\ c_{12} & c_{22} \end{vmatrix}$$

$$> 0 \quad , \ldots \begin{vmatrix} c_{11} & c_{12}, & \ldots, & c_{1n} \\ c_{12} & c_{22}, & \ldots, & c_{2n} \\ c_{1n} & c_{2n}, & \ldots, & c_{nn} \end{vmatrix} > 0 . \qquad (9.39)$$

EXAMPLE

Suppose it is desired to establish the conditions for definiteness for V having the quadratic form

$$V = \frac{1}{2} [a^2(q_1^2 + q_2^2) + 2\alpha q_3(q_1 + q_2) + b^2 q_3^2] . \qquad (9.40)$$

Since this is a quadratic of the form of equation 9.37 we can write

$$2V = c_{11}x_1^2 + 2c_{12}x_1x_2 + 2c_{13}x_1x_3$$
$$+ c_{22}x_2^2 + 2c_{23}x_2x_3 + c_{33}x_3^2 . \qquad (9.41)$$

Then

$$2V = a^2q_1^2 + a^2q_2^2 + 2\alpha q_1q_3 + 2\alpha q_2q_3 + b^2q_3^2$$

or

$$2V = a^2q_1^2 + 0 + 2\alpha q_1q_3$$
$$+ a^2q_2^2 + 2\alpha q_2q_3 + b^2q_3^2 . \qquad (9.42)$$

The discriminant

$$\Delta = \begin{vmatrix} c_{11} & c_{12} & c_{13} \\ c_{12} & c_{22} & c_{23} \\ c_{13} & c_{23} & c_{33} \end{vmatrix}$$

becomes   (9.43)

$$\Delta = \begin{vmatrix} a^2 & 0 & \alpha \\ 0 & a^2 & \alpha \\ \alpha & \alpha & b^2 \end{vmatrix} .$$

For V to be positive definite, the principal minors of the discriminant must be positive; that is,

$$a^2 > 0 , \quad \text{hence}$$
$$a^4 > 0$$

and

$$a^2(a^2b^2 - \alpha^2) + \alpha(-\alpha a^2) > 0 , \qquad (9.44)$$

or

$$a^2 > 0$$

and

$$a^2b^2 - 2\alpha^2 > 0 . \qquad (9.45)$$

9.3.2.4 *Definition of Norms*

The norm of an *n*-th dimensional vector $\vec{X}$, having components $x_1, x_2, \ldots x_n$, can be defined in the following ways:

$$\text{Euclidian norm: } \|X\| = \sqrt{X'X} = \left(\sum_{i=1}^{n} x_i^2\right)^{1/2} \quad (9.46)$$

$$\text{Taxicab norm*: } \|X\|_T = \sum_{i=1}^{n} |x_i| \quad (9.47)$$

Note that a norm or length function must satisfy the following requirements:

1. It must be zero for the zero vector;
2. it must be positive for all non-zero vectors; and
3. it must satisfy the triangle inequality, viz., if $X = X_1 + X_2$, then

$$\|X\| \leq \|X_1\| + \|X_2\| \quad (9.48)$$

By analogy, the norm of a matrix A is defined as:

$$\text{Euclidean norm: } \|A\| = \left(\sum_{i=1}^{m} \sum_{j=1}^{n} a_{ij}^2\right)^{1/2} \quad (9.49)$$

$$\text{Taxicab norm: } \|A\|_T = \sum_{i=1}^{m} \sum_{j=1}^{n} |a_{ij}| \quad (9.50)$$

9.3.3 Stability

The formal concept of stability (we shall mean stability in the sense of Liapunov) is stated, not for the system as a whole, but for one of its solutions. For example, consider the system described by

$$\dot{X} = f(X, t); X(t_o) = X_1^o \quad (9.51)$$

which has the solution $X_1(t)$ for $t \geq t_o$. Suppose now that it has the solution $(X_2(t)$ at $t > t_o$ corresponding to $X(t_o) = X_2^o \neq X_1^0$. The question arises, if X_2^o is close to X_1^o, how much does $X_2(t)$ differ from $X_1(t)$ for $t \geq t_o$? This is really a formal statement of the stability question. The stability of a system solution may thus be defined mathematically as follows:

The solution $X_1(t)$, $t \geq t_o$; $X_1(t_o) = X_1^o$ is stable if, for every parameter $\varepsilon > 0$, there exists a $\delta(\varepsilon, t_o) > 0$ such that

$$\|X_2^o - X_1^o\| < \delta(\varepsilon, t_o) ; \quad (9.52)$$

this implies that

$$\|X_2(t) - X_1(t)\| < \varepsilon, \text{ for } t \geq t_o . \quad (9.53)$$

The above definition of stability may be rephrased in the following manner. If it is possible to maintain the error in the solution to a value smaller than any desired amount, then, for all time in the future, by making the initial (condition) error small enough, the solution $X_1(t)$ is stable.

*The name derives from the measure of distance between two points used by taxicabs in a city with rectangular blocks.

In view of equations 9.52 and 9.53 the following observations can be made.

1. The statement, $X_1(t)$ is a stable solution, does not imply that $X_1(t)$ is bounded or that it tends to zero with increasing t. However, $X_1(t)$ may be identically zero in which case it is obviously bounded.
2. The concept of stability is a local one. Nothing is said about how large the initial deviations may be.
3. Either of the two norms previously defined may be used in this definition and the ones to follow. The choice of a norm is one of convenience. Often the taxicab norm is more convenient in analytical work.
4. If δ does not depend on t_o, i.e. $\delta = \delta(\varepsilon)$, then the solution is uniformly stable.

9.3.4 Asymptotic Stability

Again consider the system described by equation 9.51; namely,

$$\dot{X} = f(X, t), X(t_o) = X_1^o \quad (9.54)$$

which has the solution $X_1(t)$ for $t \geq t_o$. If the solution $X_1(t)$ is stable it will be asymptotically stable if, for another solution $X_2(t)$ (corresponding to $X_2^o \neq X_1^o$) the error, $\|X_2(t) - X_1(t)\|$, tends to zero with increasing time. That is

$$\lim_{t \to \infty} \|X_2(t) - X_1(t)\| = 0 . \quad (9.55)$$

Thus, stability in the Liapunov sense implies that the state variables will always remain within a sufficiently small neighborhood of the equilibrium Point when given a small initial perturbation. For an asymptotically stable system the state variables will always remain in a neighborhood of equilibrium and will eventually return to the equilibrium state. In contrast to this, stability in the Lagrange sense may be interpreted to mean that given a sufficiently small initial perturbation in the state variables from the equilibrium point, the norm of the difference between the state variable components and their equilibrium value will always have a finite (geometrical) bound.

The definitions of stability, asymptotic stability, and instability are illustrated in Figure 9.3.

9.4 Theory of the Direct Method of Liapunov

The fundamental concept which provides the foundation for the theory of the direct method is a simple one. It can be considered in either a geometric or a functional frame.

Consider the following two dimensional constant coefficient system of equations representing a damped oscillator:

$$\dot{x}_1 = x_2, x_1(0) = x_1^o$$
$$\dot{x}_2 = -ax_1 - bx_2, x_2(0) = x_2^o$$
$$a > 0, b > 0, 4a > b^2 \quad (9.56)$$

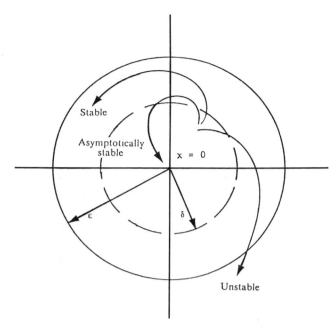

Figure 9.3 Definitions of stability in a small region near the equilibrium point.

which can be represented in a phase diagram as shown in Figure 9.4.

The coordinates of Figure 9.4 have been chosen as $\sqrt{a}x_1$ and x_2 for convenience and the direction of travel is apparent from the relation $\dot{x}_1 = x_2$. Suppose that we consider the distance squared, that is, $V = d^2$, from any point on the trajectory to the origin; therefore,

$$V = d^2 = (\sqrt{a}x_1)^2 + x_2^2 = ax_1^2 + x_2^2 . \quad (9.57)$$

The rate of change of V along the trajectory is given by

$$\frac{dV}{dt} = \dot{V} = \frac{\partial V}{\partial x_1}\frac{dx_1}{dt} + \frac{\partial V}{\partial x_2}\frac{dx_2}{dt}$$
$$= 2ax_1\dot{x}_1 + 2\dot{x}_2 x_2$$
$$= -2bx_2^2 . \quad (9.58)$$

Since $b > 0$, the distance V is always decreasing, except when $x_2 = 0$. Unless $x_2(t) \equiv 0$, $x_1(t) \neq 0$ is a solution, the trajectory must continuously tend toward the origin, as is shown in the phase diagram. If we measure the distance from the origin to a point on the trajectory and show that this distance decreases as we move along the trajectory, then the trajectory must tend toward the origin as time increases. Note that, in the example, V always decreases regardless of where we start the trajectory; thus, the solution $x_1 \equiv 0$, $x_2 \equiv 0$ is termed asymptotically stable in the large.

9.4.1 Theorems

Three theorems of the direct method define the sufficient conditions for stability, asymptotic stability, and instability of the solutions of the autonomous differential equation

$$\dot{X} = f(x), \quad X(0) = X^o, \quad f(0) = 0 . \quad (9.59)$$

9.4.1.1 Theorem I (On Stability)

If there exists a continuous scalar function V of the variables $x_1, x_2, \ldots, x_n$, i.e.,

$$V(x) = V(x_1, \ldots, x_n) \quad (9.60)$$

which is positive definite, i.e.,

$$V(x) > 0 \text{ for all } x \neq 0 \quad \text{and} \quad V(0) = 0 \quad (9.61)$$

which has first partial derivatives with respect to all the variables such that its time derivative calculated along the trajectories of equation 9.59 is nonpositive, i.e.,

$$\dot{V}(x) = \frac{\partial V}{\partial x_1}\dot{x}_1 + \ldots + \frac{\partial V}{\partial x_n}\dot{x}_n \le 0, \text{ for all } x ; \quad (9.62)$$

then the trivial solution, $x(t) \equiv 0$, of equation 9.59 is stable.

9.4.1.2 Theorem II (On Asymptotic Stability)

If there exists a Liapunov function $V(x)$ such that its time derivative calculated along the trajectories of equation 9.59 is negative definite, i.e.,

$$\dot{V}(x) = \frac{\partial V}{\partial x_1}\dot{x}_1 + \ldots + \frac{\partial V}{\partial x_n}\dot{x}_n < 0 \text{ for all } x \neq 0 ; \quad (9.63)$$

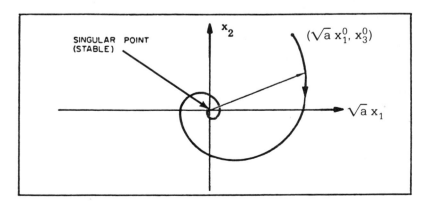

Figure 9.4 Phase diagram of a two-dimensional damped oscillator.

then the trivial solution of equation 9.59 is asymptotically stable.

9.4.1.3 Theorem III (On Instability)

If, for the differential equations of perturbed motion, one can find a function $V(x)$ whose time derivative $\dot{V}(x)$ is definite and of the same sign as $V(x)$, then the unperturbed motion is unstable.

Example I (Stability)
Consider equations of perturbed motion of the form

$$\frac{dx}{dt} = -(x - \beta t)(1 - ax^2 - by^2)$$
$$\frac{dy}{dt} = -(y - ax)(1 - ax^2 - by^2) \qquad (9.64)$$

where α, β, and a, are positive constants. For definiteness, let $a < b$, $a < \beta$ and choose a Liapunov function $V = \alpha x^2 + \beta y^2$ which is positive definite. Its total time derivative $\dot{V} = -2(\alpha x^2 + \beta y^2)(1 - ax^2 - by^2)$ is negative for sufficiently small values of x and y, hence Theorem I is satisfied and the unperturbed motion $x = 0$, $y = 0$ is stable. This does not satisfy Theorem II, however, since $\dot{V}$ is not a negative definite function [V may be zero not only when $x = y = 0$, but also when $(1 - ax^2 - by^2) = 0$].

Example II (Asymptotic Stability)
Let the equations of perturbed motion be

$$\frac{dx}{dt} = -x - y + y(x + y)$$
$$\frac{dy}{dt} = x - x(x + y) . \qquad (9.65)$$

It is desirable to check the stability of the equilibrium solution $x = y = 0$.

Choose V as $V = x^2 + y^2$; then, $\dot{V}$ becomes, when combined with equation 9.65,

$$\dot{V} = \frac{dV}{dt} = \frac{\partial V}{\partial x}\frac{dx}{dt} + \frac{\partial V}{\partial y}\frac{dy}{dt}$$
$$= 2x[-x - y + y(x + y)]$$
$$+ 2y(x - x^2 - xy) \qquad (9.66)$$
$$= -2x^2 .$$

Since V is a positive definite function, and dV/dt is a negative definite function, by Theorem II the solution $x = y = 0$ is asymptotically stable.

Example III (Instability)
Consider

$$\frac{dx}{dt} = y^3 + z^2 x$$
$$\frac{dy}{dt} = -x^3 + z^2 y$$
$$\frac{dz}{dt} = -z^3 .$$

Choose $V = x^4 + y^4$; then, the derivative $\dot{V}$ becomes

$$\frac{dV}{dt} = 4(x^4 + y^4)z^2 = 4z^2 V . \qquad (9.67)$$

Therefore, according to Theorem III, the equilibrium solution $x = y = z = 0$ is unstable since $\dot{V}(x)$ is definite and of the same sign as $V(x)$.

The sufficient conditions are stronger than the necessary conditions for stability or asymptotic stability. Although it is always possible to determine necessary and sufficient conditions for stability of the solutions to differential equations describing linear autonomous (stationary) systems, this is not so for nonlinear systems. The measure of how close the sufficient conditions are to being both necessary and sufficient is the degree of "sharpness" as it is defined in Reference 34, for example. There is, however, no known general approach for constructing a sharp Liapunov function for an arbitrary nonlinear system.

9.4.2 Criteria of Stability for the First Approximation of Perturbed Motion

Consider the equations of perturbed motion in the form

$$\frac{dx_i}{dt} = p_{i1}, x_1 + \ldots p_{in}x_n + X_i(x_1, \ldots, x_n) \qquad (9.68)$$
$$(i = 1, 2, \ldots, n)$$

where p_{ij} are constants, X_i are functions of the variables x_i which are independent of t and are resolved in the region

$$|x| \leqslant H \qquad (9.69)$$

in terms of a power series of these variables. The power series contains terms no lower than second order.

The question of the stability of the solutions of the equations 9.68 may be divided into noncritical cases in which the problem is solved by first approximation (i.e., $X_i = 0$) and the critical cases when $X_i \neq 0$. The cases are termed critical if the characteristic equation of a system of first approximation does not have roots with positive real parts but does have roots with real parts equal to zero.

The three Liapunov theorems on stability by the first approximation are now given without proof.

9.4.2.1 Theorem IV

If all the roots of a characteristic equation of a system of first approximation have negative real parts, then the unperturbed motion is stable and asymptotic as well, regardless of what the terms of the highest order in the differential equations of the perturbed motion may be.

9.4.2.2 Theorem V

If among the roots of the characteristic equation of a system of first approximation there exists at least one with a positive real part, then the unperturbed motion is unstable for any selection of terms whose order is greater than one in the differential equations of perturbed motion.

9.4.2.3 Theorem VI

If the characteristic equation of the system of first approximation has no roots with positive real parts, but has roots with positive real parts equal to zero, then the terms of a higher order may be so selected that one may obtain either stability or instability, as one chooses (critical case).

To determine if we are dealing with a critical or noncritical case the Routh-Hurwitz criteria are applicable.

Example 1
Consider a body rotating about a fixed point. Given a perturbation about this fixed point we wish to see if the motion is stable.

The differential equations of motion (Euler equations) have the following form:

$$A \frac{dp}{dt} + (C - B)qr = 0$$

$$B \frac{dq}{dt} + (A - C)rp = 0$$

$$C \frac{dr}{dt} + (B - A)pq = 0 \qquad (9.70)$$

where p, q, and r, are the vector components of the instantaneous angular velocity along the body axes of the coordinates which coincide with the principal axes of inertia of the body at the fixed point; A, B, and C are the principal moments of inertia about these axes, respectively.

Equation 9.70 has a particular solution $p = \omega =$ constant, $q = r = 0$, and two analogous solutions corresponding to two other axes of coordinates.

Now let

$$p = \omega + x$$
$$q = y$$
$$r = z \qquad (9.71)$$

where x, y, and z are small perturbations. The differential equations of perturbed motion (with constant rotation about the p axis) now become

$$A \frac{dx}{dt} + (C - B)yz = 0$$

$$B \frac{dy}{dt} + (A - C)(x + \omega)z = 0$$

$$C \frac{dz}{dt} + (B - A)(x + \omega)y = 0 . \qquad (9.72)$$

The characteristic equation of first approximation is

$$\begin{pmatrix} s & 0 & 0 \\ 0 & s & \dfrac{(A - C)}{B}\omega \\ 0 & \dfrac{(B - A)}{C}\omega & s \end{pmatrix} = 0 . \qquad (9.73)$$

Equation 9.73 has only one root equal to zero and two nonzero roots as follows:

$$s_{1,2} = \pm\,\omega\,\sqrt{\frac{(C - A)(A - B)}{CB}} . \qquad (9.74)$$

If $C < A < B$ or $C > A > B$, then both roots will be real and one of them will be positive, consequently the unperturbed motion would be unstable. Since rotation is about the A axis ($p = \omega$) and A is an intermediate axis (e.g., $C > A > B$ or $C < A < B$) this motion is unstable.

If $A < B, A < C$ or $A > B, A > C$ then both roots would be imaginary. This would then be a critical case and the first approximation is not sufficient to solve the problem. This is the case for rotation about a minor or major axis.

A general investigation of three critical roots is very complex. But in the present case the problem is solved simply as follows.

For cases $A < B < C$ or $A > B > C$ or steady rotations about the minor and major axis, respectively, the stability is obtained by considering the integral of the second and third equations in equation 9.72 which is

$$V = B(B - A)y^2 + C(C - A)z^2 = \text{constant}. \qquad (9.75)$$

Since V is a Liapunov function (positive or negative definite with $dV/dt = 0$), the system is stable by Theorem I when rotating about the major or minor axis of the body with respect to the r, q variables.

To consider stability of rotation about the middle axis of the ellipsoid of inertia, i.e., $C < A < B$ or $C > A > B$ (equations are written with constant rotation about the A moment of inertia axis) we can choose a function $V = yz$. Its derivative becomes

$$\dot{V} = \dot{y}z + \dot{z}y$$
$$= (x + \omega)\left[\frac{(C - A)}{B}z^2 + \frac{(A - B)}{C}y^2\right]. \qquad (9.76)$$

$\dot{V}$ can for $x + \omega > 0$ be of the same sign as V (positive definite when $V > 0$ or negative definite when $V < 0$) thus the configuration is unstable according to Theorem III.

9.5 Application to Automatic Control Systems

Perhaps the greatest value of the "Second Method" is in the theory of automatic control systems. There are other analytical methods for studying stability of servomechanisms with nonlinear elements such as the phase plane and describing function techniques. These, however, have some severe restrictions because the phase plane method is limited to second order time invariant differential equations, and the describing function can treat any order system but is restricted to time invariant systems with one nonlinearity. Linear systems, of course, can be treated by the Routh-Hurwitz criteria or similar methods.

The second method of Liapunov is applicable to any system with or without time dependent coefficients and with any number of nonlinear elements. A limitation of the method, however, may be the complexity of computations in systems having multiple degrees of freedom and many nonlinear elements. Other difficulties may arise due to the nonuniqueness of Liapunov functions. That is, there are many Liapunov functions which can be chosen and each will give a region of stability. These regions will not contradict each other but each will be either broader or narrower in scope than the others. Between two regions there may be a "gray" region where the stability is not well-defined.

9.5.1 Criteria for Asymptotic Stability for Some Nonlinear Systems

In the theory of servomechanisms one often meets equations of the type

$$\frac{dx_i}{dt} = d_{i1}x_1 + \ldots + d_{in}x_n + \beta_{i1}\phi_{i1}(x_1)$$
$$+ \ldots + \beta_{in}\phi_{in}(x_n) . \quad (9.77)$$
$$(i = 1, \ldots, n)$$

In studying stability one has to consider the possibility of large initial conditions (x_{io}) such that equation 9.77 may have to be considered over a large region of space (x_i). Under these conditions, the problem cannot, as a rule, be reduced to the study of a linearized system. In studying such problems, the first approximation criteria for stability are no longer sufficient without taking the nonlinearities into account. This can be done by considering the Liapunov functions which yield the required conditions for asymptotic stability.

The construction of Liapunov functions for a system such as equation 9.77 can always be obtained in the form

$$V(x, \beta, \phi) = \sum_{i,j=1}^{n} a_{ij}x_ix_j + \sum_{i,j=1}^{n} a_{ij}\int_{o}^{x_j} \phi_{ij}(\xi)d\xi \quad (9.78)$$

where the first term is a quadratic of the variables $x_1, \ldots, x_n$ and the second term is a linear combination of the integrals $\int_0^x \phi(\xi)d\xi$ of the nonlinear terms.

The time derivative dV/dt along the trajectory of equation 9.77 then determines the type of stability or instability of the system depending on its definiteness.

Example

Consider a mechanical spring mass system having one degree of freedom. The equation of motion for this system is

$$\ddot{x} + f(x, \dot{x})\dot{x} + h(x) = 0 \quad (9.79)$$

and the energy of the system is

$$V = \frac{1}{2}(\dot{x})^2 + \int_{o}^{x} h(u)du . \quad (9.80)$$

The total time derivative of equation 9.80 is, therefore,

$$\dot{V} = \dot{x}[\ddot{x} + h(x)]$$
$$= \dot{x}[-f(x, \dot{x})\dot{x} - h(x) + h(x)]$$
$$= -f(x, \dot{x})\dot{x}^2 . \quad (9.81)$$

Since $V > 0$ and $\dot{V} < 0$ will always be satisfied provided $xh(x) > 0$ and $f(x, \dot{x}) > 0$, the system is stable according to Theorem I on stability. Physically, the two restraints $xh(x) > 0$ and $f(x, \dot{x}) > 0$ imply that the spring force acts in a restoring direction and that the damping force be positive for all x and $\dot{x}$.

9.6 A General Method for the Construction of Liapunov Functions for Linear and Nonlinear Systems

The difficulty in selecting a Liapunov function is due to the fact that there is an infinite number of functions which can be considered. A systematic approach is therefore desirable which can always be relied upon to provide a suitable Liapunov function for the solution of interest. One approach, suggested in Reference 12, can be easily implemented and can be extended to nonlinear systems. This method is based on the selection of a positive definite matrix p (with elements p_{ij}) which ensures that the Liapunov function V is a quadratic form and that its derivative $\dot{V}$ can be constructed as a negative definite quadratic form thus yielding the required criteria for asymptotic stability for the solutions of the unperturbed motion. For example, consider the perturbed system of linear differential equations of the form

$$\dot{X} = AX, X(0) = X^o . \quad (9.82)$$

Choose a Liapunov function of the form

$$V = X'PX \quad (9.83)$$

then, taking the time derivative of V yields

$$\dot{V} = \dot{X}'PX + X'P\dot{X}$$
$$= X'A'PX + X'PAX$$
$$= X'(A'P + PA)X$$
$$= -X'QX . \quad (9.84)$$

If P is positive definite then $V > 0$ for $X \neq 0$ and $V \to \infty$ as $\|X\| \to \infty$. If Q is also positive definite, then $\dot{V} < 0$ for $X \neq 0$ and the system described by equation 9.82 is completely stable (stable in the large) if solution P of

$$A'P + PA = -Q \quad (9.85)$$

is positive definite.

With no loss of generality, let P and Q be symmetric matrices since

$$-Q' = (A'P + PA)'$$
$$= P'A + A'P'$$
$$= A'P + PA = -Q . \quad (9.86)$$

In this case, the n^2 equations of equation 9.85 reduce to $n(n + 1)/2$ equations because $p_{ij} = p_{ji}$. Thus, the stability is determined by (1) choosing any positive definite Q (such as a unit matric I, for example), (2) computing p_{ij} from $n(n + 1)/2$ linear algebraic equations and, finally, (3) testing for positive definiteness by Sylvester criterion.

Example
Consider

$$\frac{d^2x}{dt^2} + a_1 \frac{dx}{dt} + a_2 x = 0 \qquad (9.87)$$

or, in the first order form,

$$\dot{x} = y - a_1 x$$
$$\dot{y} = -a_2 x$$

or

$$\begin{pmatrix} \dot{x} \\ \dot{y} \end{pmatrix} = \begin{pmatrix} -a_1 & 1 \\ -a_2 & 0 \end{pmatrix} \begin{pmatrix} x \\ y \end{pmatrix} = AX \qquad (9.88)$$

where

$$X = \begin{pmatrix} x \\ y \end{pmatrix} \qquad A = \begin{pmatrix} -a_1 & 1 \\ -a_2 & 0 \end{pmatrix}.$$

Assume $Q = I =$ unit matrix (positive definite), then equation 9.85 becomes

$$\begin{pmatrix} -a_1 & -a_2 \\ 1 & 0 \end{pmatrix} \begin{pmatrix} p_{11} & p_{12} \\ p_{12} & p_{22} \end{pmatrix}$$
$$+ \begin{pmatrix} p_{11} & p_{12} \\ p_{12} & p_{22} \end{pmatrix} \begin{pmatrix} -a_1 & 1 \\ -a_2 & 0 \end{pmatrix} = - \begin{pmatrix} 1 & 0 \\ 0 & 1 \end{pmatrix}$$

which yields

$$p_{11} = \frac{a_2 + 1}{2a_1}$$

$$p_{12} = -\frac{1}{2}$$

and

$$p_{22} = \frac{(a_2 + 1 + a^2{}_1)}{2a_1 a_2}.$$

For P to be positive definite by Sylvester's theorem $p_{11} > 0$ and

$$\begin{vmatrix} p_{11} & p_{12} \\ p_{12} & p_{22} \end{vmatrix} = p_{11} p_{22} - p_{12}^2 > 0$$

or

$$\frac{a_2 + 1}{2a_1} > 0$$

and

$$(a_2 + 1) \frac{(a_2 + 1 + a_1^2)}{4a_1^2 a_2} - \frac{1}{4}$$
$$= \frac{(a_2 + 1)^2 + a_1^2}{4a_1^2 a_2} > 0 . \qquad (9.89)$$

This implies that $a_2 > 0$ and then from $(a_2 + 1)/2a_1 > 0$ it follows that $a_1 > 0$. The conditions $a_1 > 0$ and $a_2 > 0$ are the Routh-Hurwitz criteria of sufficient (but nonnecesary) stability for the given differential equation.

9.6.1 Extension to Nonlinear Systems

Consider now perturbed equations for the solutions of a nonlinear system in the form [12]

$$\dot{X} = AX + g(X), \, g(0) = 0 \qquad (9.90)$$

or when expanded

$$\begin{pmatrix} x_1 \\ x_2 \\ \cdots \\ x_n \end{pmatrix} = \begin{pmatrix} a_{11} & a_{12} & \cdots \\ \cdots & \cdots & \cdots \\ \cdots & \cdots & \cdots \\ \cdots & \cdots & \cdots \end{pmatrix} \begin{pmatrix} x_1 \\ x_2 \\ \cdots \\ x_n \end{pmatrix} + \begin{pmatrix} g_1 \\ g_2 \\ \cdots \\ g_n \end{pmatrix} \qquad (9.91)$$

where $g(X)$ contains only quadratic and higher order terms. Also assume that A is a stable matrix (all roots have negative real parts) which implies that the origin of the phase space is asymptotically stable. Then again let

$$V = X'PX \qquad (9.92)$$

therefore,

$$\dot{V} = \dot{X}'PX + X'PX'$$
$$= (AX + g)'PX + X'P\dot{X}$$
$$= (X'A' + g')PX + X'P(AX + g)$$
$$= X'A'PX + g'PX + X'PAX + X'Pg$$
$$= X'(A'P + PA)X + g'PX + (g'P'X)' . \qquad (9.93)$$

Now, $(g'P'X)'$ is a scalar (transpose) $= X'Pg$, and $g'PX = $ scalar $=$ transpose $= X'P'g \equiv X'Pg$.

Also, P was assumed to be symmetric, therefore

$$\dot{V} = X'(A'P + PA)X + 2X'Pg . \qquad (9.94)$$

Hence the origin of the linear system $\dot{X} = AX$ is asymptotically stable when

$$A'P + PA = -Q = I = \text{unit matrix} \qquad (9.95)$$

Since Q can be any positive definite matrix, then inside the domain

$$\dot{V} = -X'QX + 2X'Pg$$
$$= -X'X + 2X'Pg \qquad (9.96)$$

where P is positive definite, the nonlinear system equation 9.90 will also be asymptotically stable. The boundary is found by solving for $\dot{V} = 0$ or

$$X'X = 2X'Pg . \qquad (9.97)$$

The stability region can thus be obtained for the nonlinear system without integration of the original equations. The P matrix can be chosen for convenience and experience with the problem as an arbitrary positive definite matrix. The choice of optimum P is not generally known.

9.7 Synthesis of a Control System by the Direct Method

Consider the following example where it is desired to determine the control torques which will bring a freely spinning body to rest. Assuming that there are three independent sources of torques in the body, Euler's equations are of the form:

$$I_1\dot{\omega}_1 = (I_2 - I_3)\omega_2\omega_3 + u_1$$
$$I_2\dot{\omega}_2 = (I_3 - I_1)\omega_3\omega_1 + u_2$$
$$I_3\dot{\omega}_3 = (I_1 - I_2)\omega_1\omega_2 + u_3 . \qquad (9.98)$$

Here, $u_i(i = 1 - 3)$ are control torques. It is necessary to choose u_i as functions of the angular rates ω_i and inertias I_i in order to reduce ω_i, in time, to zero. Trying

$$V(\omega) = \frac{1}{2}(I_1\omega_1^2 + I_2\omega_2^2 + I_3\omega_3^2)$$
$$= \text{kinetic energy} \qquad (9.99)$$

then

$$\dot{V} = \omega_1 u_1 + \omega_2 u_2 + \omega_3 u_3 \qquad (9.100)$$

to a first approximation. If the body comes to rest, then $V \to 0$ and $\dot{V} < 0$. Therefore, imposing the constraint

$$\dot{V} = -\alpha V \qquad (9.101)$$

leads to

$$\omega_1 u_1 + \omega_2 u_2 + \omega_3 u_3 = \qquad (9.102)$$
$$-\frac{\alpha}{2}(I_1\omega_1^2 + I_2\omega_2^2 + I_3\omega_3^2)$$

with solutions

$$u_i = -\frac{\alpha}{2} I_i \omega_i \qquad (i = 1, 2, 3) . \qquad (9.103)$$

Integrating equation 9.100 yields

$$V[\omega(t)] = V[\omega(0)]e^{-\alpha t} . \qquad (9.104)$$

Since $V(0) = 0$, $V(\omega) > 0$ for all $\omega \neq 0$, and $V(\omega) \approx \omega^2$ we see that with the linear control $\omega_i \to 0$ asymptotically with the approximate time constant $2/\alpha$. Thus, the highly nonlinear system with a linear control law (9.103) behaves somewhat like a linear system. Equation 9.104 can be used to compute time from some $\omega'(t_o)$ to some $\omega''(t)$, i.e.,

$$V[\omega''(t)] = V[\omega'(t_o)]e^{-\alpha(t - t_o)} \qquad (9.105)$$

and

$$T = t - t_o = \frac{1}{\alpha} ln \frac{V(\omega')}{V(\omega'')} . \qquad (9.106)$$

9.8 Stability Analysis Example

In addition to the classical linearization methods involving the characteristic equation of a system of constant coefficient equations, Liapunov's direct or second method may often be used to facilitate stability analysis. The main difficulty in the application of Liapunov's techniques has been in finding a relatively simple Liapunov function which still has physical meaning. For a mechanical system, where the coordinate constraints are scleronomic (free from explicit time dependence), this difficulty can be avoided by selecting the Hamiltonian as the Liapunov function [16]. The Hamiltonian approach is particularly well adapted for use in connection with the Lagrangian formulation when it is used to derive the system's rotational equations of motion.

An example of a Hamiltonian used as a Liapunov function to examine the stability of a gravity gradient stabilized satellite in geosynchronous orbit is given in reference 15. The approach is as follows:

Consider the gravity gradient configuration in Figure 9.5.

The three axis systems employed to describe the orientation of the two bodies relative to each other and to the local vertical reference are:

1. ξ, η, ζ–main body principal axes
2. d–damper boom axis
3. X, Y, Z–local vertical system axes

Lagrange's general equations of motion applicable to a nonconservative system with linear viscous damping are of the form

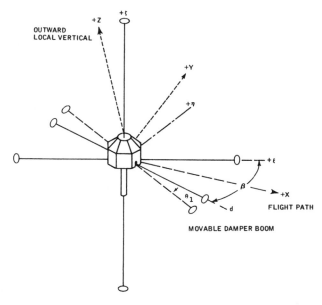

Figure 9.5 Gravity gradient configuration (DODGE) [15].

$$\frac{d}{dt}\left(\frac{\partial L}{\partial \dot{q}_i}\right) - \frac{\partial L}{\partial q_i} = -\frac{\partial F}{\partial \dot{q}_i},$$

$$i = 1, \ldots, N \qquad (9.107)$$

N degrees of freedom

where T and V represent the total (complete) rotational kinetic and potential energies of the system respectively and $L = T - V$.

The generalized coordinates q_i and velocities $\dot{q}_i$ selected here are

$q_1 = \varphi$–pitch degree of freedom, $\dot{q}_1 = \dot{\varphi}$
$q_2 = \theta$–roll degree of freedom, $\dot{q}_2 = \dot{\theta}$
$q_3 = \psi$–yaw degree of freedom, $\dot{q}_3 = \dot{\psi}$
$q_4 = \theta_1$–damper boom rotational degree
    of freedom, $\dot{q}_4 = \dot{\theta}_1$

Also, F = the Rayleigh dissipation function.

The damping is assumed to be proportional to the relative angular motion, θ_1, between the satellite main body and the damper boom. The Rayleigh dissipation function may be used for nonconservative systems involving only equivalent linear viscous damping [17].

The Rayleigh function $F(\dot{q}_i)$ is explicitly a function of the generalized velocities, $\dot{q}_i$, such that the resulting dissipative forces Q_i are derivable, treating F as a 'potential type' term; namely,

$$Q_i = -\frac{\partial F}{\partial \dot{q}_i}. \qquad (9.108)$$

For this application,

$$F = \frac{1}{2}\,\omega k \dot{\theta}_1^2 \qquad (9.109)$$

yielding the dissipative damping force

$$Q_\theta = -k\omega\dot{\theta}_1 \qquad (9.110)$$

where k is the dimensionless torsional (equivalent viscous) damping constant.

Lagrange's general equations of motion (equation 9.106) represent a system of N second order differential equations which may also be written in the form of Hamilton's canonical equations as follows:

$$\dot{p}_i = \frac{\partial H}{\partial q_i} + Q_i$$

$$\dot{q}_i = \frac{\partial H}{\partial p_i} \qquad (9.111)$$

Here $H = H(p, q, t)$ is the Hamiltonian function and $Q_i = -\partial F/\partial \dot{q}_i$. Equation 9.111 thus represents a system of $2N$ first order differential equations, where the Hamiltonian is related to the Lagrangian by

$$H = \sum_{i=1}^{N} (p_i\dot{q}_i - L) \qquad (9.112)$$

with $L = T - V$ and $p_i = \partial L/\partial \dot{q}_i$ for $i = 1$ to N.

The total time derivative of the Hamiltonian is given by:

$$\frac{dH}{dt} = \frac{\partial H}{\partial t} + \sum_{i=1}^{N} \frac{\partial H}{\partial q_i}\dot{q}_i + \frac{\partial H}{\partial p_i}\dot{p}_i \qquad (9.113)$$

which simplifies to

$$\frac{dH}{dt} = Q = \sum_{i=1}^{N} Q_i\dot{q}_i. \qquad (9.114)$$

Here H is not explicitly time dependent, as in the case of scleronomic coordinate constraints.

In general, the kinetic energy may be expressed as [16, 18]

$$T = T_o + T_1 + T_2 \qquad (9.115)$$

with T_2 a quadratic form in the generalized velocities, T_1 a linear form in these derivatives, and T_o independent of them. After substitution of $T - V$ for L and $T_o + T_1 + T_2$ for T into equation 9.112, the Hamiltonian becomes

$$H = T_2 + V - T_o. \qquad (9.116)$$

Both V and T_o are independent of the generalized velocities, and when T_o does not involve time explicitly, then the quantity $V - T_o$ may be grouped together as a scalar function of the coordinates. That is,

$$P = V - T_o \qquad (9.117)$$

where P is called the dynamic potential.

For this application, the total time derivative of the Hamiltonian is expressed as

$$Q = Q_{\theta_1}\dot{\theta}_1 = -\omega I_o k \dot{\theta}_1^2 \qquad (9.118)$$

where k, the dimensionless damping constant, is always positive. For this completely damped mechanical system dH/dt is identically zero for all-time if and only if the variational coordinates are all zero simultaneously (i.e., the system is in equilibrium). Thus the motion "passes through" the instantaneous time points where $\dot{\theta}_1$ is zero, and H decreases over a finite interval of time surrounding this point.

The Hamiltonian in equation 9.116 is given as $H = T_2 + P$ where the T_2 function is always positive definite in a neighborhood of the zero solution. Thus, the sufficient condition for stability is the positive definiteness of the dynamic potential,

$$P = V - T_o = V_b + V_d + V_{\theta_1} - T \qquad (9.119)$$

If P is examined at near equilibrium conditions where ϕ, θ, θ_1 are small and where $\psi \to \psi_o$ and $\psi_1 \to \psi_{1_o}$, then $2P$ becomes a homogeneous, quadratic form in the generalized coordinates and may be tested for positive definiteness using Sylvester's criterion as follows. Applying Sylvester's criterion, it is seen that the principal nontrivial condition for stability is the positiveness of the corresponding fourth order determinant whose elements are composed of the second order partial derivatives of $2P$ with respect to the generalized coordinates. The following inequality results

$$\frac{K - 3 - S^2\psi_{1_o}}{I_o} > \frac{4S^2\psi_{1_o}}{I_\xi S^2\psi_o + I_\eta C^2\psi_o - I_\zeta - I_o S^2\psi_{1_o}}$$
$$+ \frac{3C^2\psi_{1_o}}{I_\xi C^2\psi_o + I_\eta S^2\psi_o - I_o C^2\psi_{1_o} - I_\zeta} \tag{9.120}$$

as the principal (nontrivial) sufficient condition for the positive definiteness of the dynamic potential and, thus, the Hamiltonian.

Here S represents sine and C the cosine function of the main body yaw angle (ψ_o) and the damper boom angle (ψ_{1_o}). K is a dimensionless torsional constant of the damper gimbal, and I_ξ, I_η, I_ζ are the main body principal axes. I_o is the moment of inertia of the damper boom about its hinge axis.

The asymptotic stability of the system is dependent upon the satisfaction of inequality (equation 9.120) and also upon the presence of damping, but not on the magnitude of damping (i.e., there is no restriction on the magnitude of Q).

Reference 19 reviews the theory of stability in a historical perspective and provides an extensive bibliography on the subject.

9.9   References

1. Szebehely, V. "Review of Concepts of Stability," Celestial Mechanics, Vol. 34, 1984, pp. 49–64.

2. LaSalle, J., and S. Lefschetz. *Stability by Liapunov's Direct Method*, Academic Press, Inc., New York, 1961.

3. Takahashi, Y., Rabins, M. J., and D. M. Auslander. *Control and Dynamic Systems*, Addison-Wesley, 1970.

4. Hahn, W. *Theory and Application of Liapunov's Direct Method*, Prentice-Hall, Inc., Englewood Cliffs, NJ, 1963.

5. Hayashi, C. *Nonlinear Oscillations in Physical Systems*, McGraw-Hill, New York, 1960.

6. Beletskii, V. V. *Motion of an Artificial Satellite About Its Center of Mass*, NASA Translat., 1966.

7. Hughes, P. C. *Spacecraft Attitude Dynamics*, John Wiley & Sons, 1986.

8. Kalman, R. E., and J. E. Bertram. "Control System Analysis and Design via the 'Second Method' of Liapunov," Journal of Basic Engineering, ASME Transaction, Vo. 82, No. 2, pp. 371–400, June 1960.

9. Routh, E. J. *Stability of Motion*, Taylor and Francis, London, 1975.

10. Pipes, L. A. *Applied Mathematics for Engineers and Physicists*, McGraw-Hill, 1958.

11. Mingori, D. L. "Stability Analysis of Dual-Spin Satellites," The Aerospace Corporation, TR-0158(3133–02)-1, July 1967.

12. Ince, E. L. *Ordinary Differential Equations*, Dover, NY, 1956.

13. Richards, J. A. *Analysis of Periodically Time Varying Systems*, Springer-Verlag, New York, 1983.

14. Likins, P. "Linearization and Liapunov Stability Analysis of a Class of Dynamical Differential Equations," AIAA Journal, Vol. 10, April 1972, pp. 453–456.

15. Bainum, P. M., and D. L. Mackison. "Gravity-Gradient Stabilization of Synchronous Orbiting Satellite," Journal of the British Interplanetary Society, Vol. 21, pp. 341–369, 1968.

16. Pringle, R., Jr. "On the Capture, Stability, and Passive Damping of Artificial Satellites," Ph.D. dissertation, Stanford University; also, NASA Report No. CR-129, December 1964.

17. Goldstein, H. *Classical Mechanics*, Addison-Wesley Publishing Co., 1959.

18. Marandi, S. R., and V. J. Modi. "Use of the Energy Function in Determining Stability Bounds of Asymmetric Satellites," J. Astronautical Sciences, Vol. 37, No. 2, April-June 1989, pp. 121–143.

19. Pradeep, S., and S. K. Shrivastava. "Stability of Dynamical Systems: An Overview," J. Guidance, Control and Dynamics, May-June 1990, pp. 385–393.

Appendix A

Problem Sets

A.1 Problem Set 1: Kinematics

1. A bar $AB = 3$ m is moving in a plane. At a given instant as shown in Figure A.1 the velocity of A and B are:

   $V_A = 2$ m/s acting 60 degrees clockwise from line A to B

   V_B acting 30 degrees clockwise from line A to B

   Determine the angular velocity ω of the bar stating whether it is clockwise or counterclockwise.

   (Ans. $\omega = 0.383$ rad/s counterclockwise).

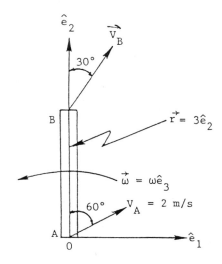

Figure A.1

2. Show that the angular deviation due to the Coriolis acceleration for a particle moving away from the north pole with a constant velocity v is 2π radians per day measured with respect to the Earth's surface.

3. Determine the deviation of a falling body on the Earth's surface from a height h at latitude λ.

   (Ans. $\dfrac{\omega}{3}\sqrt{\dfrac{8h^3}{g}}\cos\lambda$, east of vertical)

4. At north latitude λ a projectile is fired in the vertical plane of a plumb line at an elevation α in the easterly direction with velocity v. Determine the Coriolis acceleration of the projectile as seen by an observer on the Earth.

   (Ans. $a_c = 2\omega v(\sin\lambda\cos\alpha\hat{e}_1 - \cos\lambda\sin\alpha\hat{e}_2 + \cos\lambda\cos\alpha\hat{e}_3)$

5. Determine the velocity and acceleration of a point $P(r_1, r_2, r_3)$ which is fixed in a body rotating at a constant angular rotation rate $\vec{\omega}$. Assume $\vec{\omega} = \omega_1\hat{e}_1 + \omega_2\hat{e}_2 + \omega_3\hat{e}_3$ and express the results in the body reference frame.

6. Find the magnitude and direction of the centripetal acceleration $\vec{\omega} \times (\vec{\omega} \times \vec{r})$ for the particle moving in a circular path of radius a with a constant angular velocity ω as shown in Figure A.2.

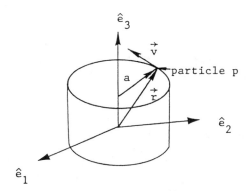

Figure A.2

7. A particle of propellant moves radially outward from a fuel tank located on a spacecraft with a constant radial velocity $v_r = 0.1$ m/s. The body is rotating with a uniform counterclockwise angular velocity of 60 revolutions per minute. What is the acceleration of the particle at the point where it leaves the body at a radial distance of 1 m?

   (Ans. 39.5 m/s²)

8. Show that the rotation matrix

$$R = \begin{pmatrix} 1 & 0 & 0 \\ 0 & \cos\theta & \sin\theta \\ 0 & -\sin\theta & \cos\theta \end{pmatrix}$$

is orthogonal (i.e., $R^{-1} = R'$).

9. Determine the inertial components of a point $(1, 3, 4)$ in the rotating coordinate system $\hat{e}_1, \hat{e}_2, \hat{e}_3$, specified by the Euler angles $\theta = 10°$, $\psi = 20°$, and $\phi = 30°$.

   (Ans. -1.40, 1.99, 4.48)

10. Derive the body angular velocity components $\omega_1, \omega_2, \omega_3$ in terms of the Euler angular rates $\dot{\psi}, \dot{\theta}, \dot{\phi}$.

11. Express the Euler angular rates $\dot{\psi}, \dot{\theta}, \dot{\phi}$ in terms of the body velocity components $\omega_1, \omega_2, \omega_3$.

12. Find the associated Euler angles from the R_{123} rotation matrix. What can be done when $\theta_2 = 90°$?

13. Find the associated Euler angles for the case of the classical Euler angle sequence represented by equation 1.16.

14. (a) For a classical Euler sequence of rotations, as illustrated in Figure 1.5, the rotation matrix R is given by equation 1.14. Compute the elements of R if $\psi = 45°$, $\theta = 45°$, and $\phi = 30°$.

    (b) Using the definition of quaternions in equations 1.36 compute $q_i(i = 1 - 4)$.

       (Ans. $q_1 = 0.379$, $q_2 = 0.049$, $q_3 = 0.562$, $q_r = 0.733$)

    (c) Compute Euler axis (eigenvector λ) and its components m_1, m_2, m_3 along the $\hat{E}_1, \hat{E}_2$, and $\hat{E}_3$ unit vectors. Compute the Euler rotation angle μ about λ.

       (Ans. $m_1 = 0.558$, $m_2 = 0.073$, $m_3 = 0.827$, $\mu = 85.7°$)

    (d) Extract θ, ϕ, ψ from the rotation matrix R.

15. A rigid body whose principal axes (e_1, e_2, e_3) are initially aligned with the reference axes (E_1, E_2, E_3) undergoes a rotation such that the e_1, e_2, e_3 axes are aligned with the $-E_2, E_3, -E_1$ axes, respectively, Compute the Euler axis (eigenvector λ) and the Euler rotation angle μ about the Euler axis.

    (Ans. $m_1 = 1/\sqrt{3}$, $m_2 = -1/\sqrt{3}$, $m_3 = -1/\sqrt{3}$, $\mu = 120°$)

16. For the case of body axes (e_1, e_2, e_3) initially aligned with the reference axes (E_1, E_2, E_3) determine the quaternion components q_i $(i = 1 - 4)$ and the rotation matrix R for each sequential 90° rotation of the body axes about the E_3, rotated e_2, and rotated e_1 axes. What is the final rotation matrix R?

17. Assume that the body axes e_1, e_2, e_3 are initially aligned with the reference axes E_1, E_2, E_3. If the body undergoes a 90° rotation about the E_3 axis, and subsequently another 90° rotation about the rotated e_2 axis, find the following:
    (a) The rotation matrices R_1 and R_2 with the corresponding quaternions Q_1 and Q_2 for each rotation separately.
    (b) Final (combined) rotation matrix R_{12} and the corresponding quaternion Q_{12}.
    Verify the results in part (b) by the combined rotation method of equation 1.53.

18. Verify the properties of the direction cosines given in the following:
    (a) equation 1.22.
    (b) equation 1.23.
    (c) equation 1.24.

19. Show that the expanded form of equation 1.31 is given by equation 1.33.

A.2 Problem Set 2: Dynamics

1. Three particles of masses 2, 3, and 4 kg are located at the points 1, 2, 4 along the e_1, e_2, e_3 axes measured in meters. Find the moments of inertia about these axes.

   (Ans. $I_1 = 76$ kg $\cdot$ m^2, $I_2 = 66$ kg $\cdot$ m^2, $I_3 = 14$ kg $\cdot$ m^2)

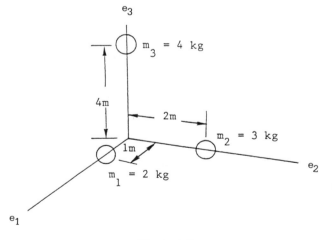

Figure A.3

2. Find the moment of inertia of a uniform thin rod of mass M and length $2b$ about an axis perpendicular to the rod (a) passing through the center, and (b) passing through one end.

   $\left(\text{Ans. (a) } I = \dfrac{Mb^2}{3}; \text{ (b) } I = \dfrac{4Mb^2}{3}\right)$

3. Find the moment of inertia of a uniform thin disk of mass M and radius a about (a) a centroidal axis normal to the disk and (b) its diagonal.

$$\left(\text{Ans. (a) } I = \frac{Ma^2}{2} \text{ , (b) } I = \frac{Ma^2}{4} \right)$$

4. Two particles of masses m_1 and m_2, respectively, are connected by a rigid massless rod of length ℓ. Show that the moment of inertia of the system about an axis perpendicular to the rod and passing through the center of mass is $m\ell^2$ where the reduced mass is $m = m_1 m_2/(m_1 + m_2)$.

5. Find the moment of inertia of a homogeneous rectangular plate of mass M with the edges 2a and 2b about an axis passing through the center and which is parallel to edge 2a.

$$\left(\text{Ans. } I = \frac{Mb^2}{2} \right)$$

6. Find the moment of inertia of a homogeneous rectangular block of mass M with edges of length 2a, 2b, and 2c about an axis passing through the center and which is parallel to edge of length 2a.

$$\left(\text{Ans. } I = \frac{M}{3} (b^2 + c^2) \right)$$

7. Derive the moments of inertia for a homogeneous solid cylinder of radius R, height H, and mass M with respect to its centroidal axis of symmetry.

$$\left(\text{Ans. } I_1 = I_2 = \frac{M}{12} (3R^2 + H^2), I_3 = \frac{MR^2}{2} \right)$$

8. Derive an expression for the moment of inertia of mass for a solid homogeneous sphere having a radius R, with respect to a diameter of the sphere.

$$\left(\text{Ans. } I_1 = I_2 = I_3 = \frac{2MR^2}{5} \right)$$

9. Show that if A, B, and C are principal moments of inertia of a body, then $A + B > C$.

10. Derive the moments and products of inertia for a homogeneous thin square plate of length "a" with respect to its edges (see sketch). Find the principal moments of inertia and the principal axes.

$$\left(\text{Ans. } I_{11} = I_{22} = \frac{ma^2}{3} \text{ , } I_{33} = \frac{2ma^2}{3}, I_{23} = I_{32} = \frac{-ma^2}{4} \right)$$

$$\left(\begin{matrix} \text{Principal moments} \\ \text{of inertia:} \end{matrix} \quad I_1 = \frac{ma^2}{12}, I_2 = \frac{7ma^2}{12}, I_3 = \frac{2ma^2}{3} \right)$$

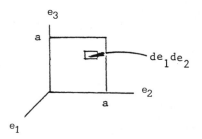

Figure A.4

11. A thin disk is spun about an axis making an angle γ with the normal to the disk and then released. Assuming the disk to be moment free, find the half angle θ of the cone generated by the normal to the disk about its angular momentum vector and the time required for the normal to make one complete rotation around the cone. What is the period of rotation if γ is small? (Hint: for a disk $C = 2A$ where A is the transverse moment of inertia of the disk).

$$\left(\text{Ans. } P = 2 \pi/\dot\psi \approx \frac{\pi}{\omega} \text{ for } \gamma \text{ small} \right)$$

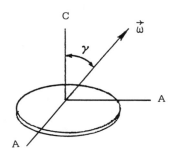

Figure A.5

12. Two equal masses connected by a massless rigid rod (a dumbbell) are rotating about an axis which is at an angle θ to the rod (see Figure A.6). Find the kinetic energy, the angular momentum, and the torque being applied to the system. The angular velocity ω is constant.

(Ans. $\vec{h} = 2ma^2\omega \sin \theta \hat{e}_3$,   K. E. $= ma^2\omega^2 \sin^2 \theta$

   Torque $= 2ma^2\omega^2 \sin \theta \cos \theta$)

13. At a given instant the angular momentum of a body about the center of mass is $\vec{h} = 15\hat{e}_1 + 10\hat{e}_2 + 12\hat{e}_3$ N · m · s. The inertia matrix about the same coordinate axes is

$$I = \begin{pmatrix} 50 & 0 & 0 \\ 0 & 40 & -20 \\ 0 & -20 & 30 \end{pmatrix} \text{ kg} \cdot \text{m}^2$$

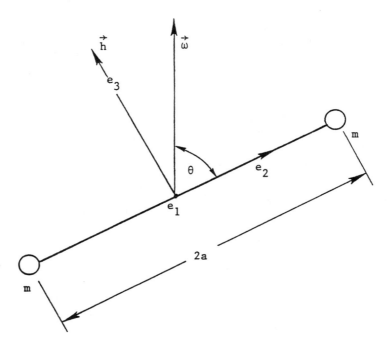

Figure A.6

Find the angular velocity and the kinetic energy of the body at the given instant. (Hint solve equation 1.65 for ω_1, ω_2, ω_3.)

(Ans. $\vec{\omega} = 0.3\hat{e}_1 + 0.675\hat{e}_2 + 0.85\hat{e}_3$ rad/s, $T = 10.725$ N · m)

14. A thin bar of mass m and length ℓ is pinned to the e_2 axis at a distance "a" from the origin and rotates about the e_1' asix, parallel to the e_1 axis, at speed $\dot{\theta}$. (See Figure A.7.) If the e_1, e_2, e_3 coordinates rotate about the e_3 axis with angular speed Ω, determine the e_1, e_2, and e_3 components of the angular momentum of the bar.

$$\left(\text{Ans. } h_1 = m\ell^2\dot{\theta}/12, \ h_2 = \frac{m\ell^2}{12}\sin\theta\cos\theta, \right.$$

$$\left. h_3 = \Omega\left(ma^2 + \frac{m\ell^2}{12}\cos^2\theta\right)\right)$$

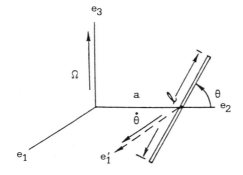

Figure A.7

15. A symmetric body with components of inertia A, A, C is damped with torques about the body fixed axes as follows: $T_1 = -k\omega_1$, $T_2 = -k\omega_2$, $T_3 = -k\omega_3$. Show that the angle between the spin axis (axis 3) and the angular momentum vector is

$$\tan\theta = \frac{|\omega_{12}(0)|}{\omega_3(0)} \cdot \frac{A}{C}\exp\left[-\left(\frac{C}{A} - 1\right)\frac{kt}{C}\right]$$

where $\omega_{12}(0) = [\omega_1^2(0) + \omega_2^2(0)]^{1/2}$ = initial transverse component of angular velocity.

16. The scout vehicle fourth stage has a radius $R = 1/3$ m and the longitudinal (spin) axis moment of inertia $C = 30$ kg · m^2. If four rocket motors of 340 N · s impulse each are used to spin up the stage then calculate the following:

(a) Maximum spin rate ω_s of fourth stage and the spin axis angular momentum H_s.

(Ans. 144.3 rpm, 453.3 N · m · s)

(b) Nutation half angle θ and $\dot{\psi}$ rate (torque-free precession) if the transverse moment of inertia $A = 8C$ and angular momentum component $\Delta H = H_s/100$.

(Ans. 0.573°, $\dot{\psi} = 18$ rpm)

(c) Transverse angular velocity ω_t and body cone angle γ.

(Ans. 1.08°/s, 0.0715°)

17. Assume that the longitudinal (spin axis) component of the spacecraft angular velocity is 60 rpm. Determine the following for an initial nutation angle of 10 degrees and spin and transverse moments of inertia of 700 and 2000 kg, respectively:
    (a) Magnitude of angular velocity ω.
    (b) Wobble angle γ.
    (c) Torque-free precession rate $\dot{\psi}$.
    (d) Nutation frequency λ.

    (Ans. 0.00646 rad/s, 3.36°, 0.355 rpm, −0.068 rad/s)

18. The impulse from a radially mounted thruster on a spinning spacecraft is delivered over some fraction of a revolution. Show that if the thruster firing starts at an angle $\Delta\phi/2$ away from the desired thrust direction and stops at the same angle past the desired direction that the firing efficiency is $\sin(\Delta\phi/2)/\Delta\phi/2$. Here $\Delta\phi$ is the angle over which the thruster firing occurs as shown in Fig. A.8.

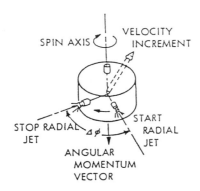

Figure A.8

19. A rocket/spacecraft configuration in Figure 2.14a has the transverse and spin axis moments of inertia before separation of $A = 80$ and $C = 5$ kg · m², respectively. After separation they are $A_1 = 20$, $A_2 = 50$, $C_1 = 1$, $C_2 = 4$ kg · m², respectively. Assuming that the spin axis and transverse angular velocity components before separation are 2 and 0.1 rad/s, respectively, compute:
    (a) Nutation angles before and after separation.

    (Ans. $\theta_1 = 17.7°$, $\theta_2 = 11.3°$)

    (b) Torque-free precession rates before and after separation.

    (Ans. $\dot{\psi}_b = 0.393$ rad/s, $\dot{\psi}_{a_1} = 0.502$ rad/s, $\dot{\psi}_{a_2} = 0.314$ rad/s)

20. A space station is hit by an object and air begins to escape. What is the final spin speed (angular rate) of the station after an hour if the initial rate was zero and the tangential torque from the escaping air is of the form:

$$T = Ke^{-\alpha t}$$

where $K = 1000$ N · m, $\alpha = 1 \times 10^{-4}$ rad/s and the spin moment of inertia $I = 1 \times 10^6$ kg · m²?

    (Ans. 20 rpm)

21. Compute the propellant mass required to:
    (a) Spin up a satellite to 60 rpm.
    (b) Reorient (precess) the satellite 5°.
    Assume the following:

    1. Spin inertia $= 700$ kg · m²
    2. Rocket thrust $F = 2.5$ N
    3. Spin moment arm $R = 1$m
    4. Axial thruster moment arm $R = 1$ m
    5. Pulse duration $\Delta t = 0.25$ sec (1/4 of spin period)
    6. Specific impulse $I_{sp} = 200$ s

    (Ans. 2.24 kg, 0.217 kg)

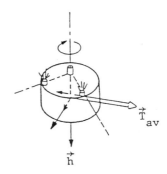

Figure A.9

22. Calculate propellant mass required and the maneuver duration to slew a spinning spacecraft through 360°. Assume the following:

    Spin inertia $= I = 1200$ kg · m².
    Angular momentum $H = I\omega = 627.44$ N · m · s.
    Thruster moment arm $R = 1.2$ m.
    Angular firing interval $\Delta\phi = 90°$.
    Specific impulse $I_{sp} = 235$ s.
    Thrust force $F = 10$ N.

23. Assume that the energy dissipation rate for a sloshing half-filled spherical tank of fluid (propellant) is $\dot{T} = -10^{-3}\omega_t^2$(N · m)/sec where $\omega_t =$ transverse angular rate of a dual-spin spacecraft due to nutation. Find the dedamping (divergence) time constant (ω_t as a function of time) for an initial nutation angle θ_o. Use equation 3.14 in text with $\dot{T}_1 = 0$ (no energy dissipation on platform). Assume:

$\omega_3 = 0$

$\Omega = 60$ rpm = spin rate of spinner

$C_2 = 130$ kg · m² = rotor spin inertia

$A = 150$ kg · m² = configuration transverse inertia

(Ans. 6.39 hr)

24. Derive the equation for the precession of a top (gyroscope subject to constant gravitational torque). What is the precession rate when the nutation angle $\theta = 90°$?

25. (a) What is the primary function of a gyroscope? What are the two properties of a spinning mass that provide the primary function of a gyroscope?
    (b) If the gyroscope is at the equator with its spin axis pointing north, where will the spin axis be pointing 12 hours later? If the spin axis were pointing up (away from the Earth), where would it be pointing 12 hours later?
    (c) Given: a gyro has $I = 10$ kg · m², $\Omega = 20$ rad/s. What is the precession rate if a constant 100 N · m torque acts normal to the spin vector?

26. It is desired to completely despin a spacecraft by using a yo-yo device. Compute the required cord length and time assuming tangential and radial release of yo-yo masses. Use the following constants and parameters:

Spacecraft spin speed   = 60 rpm.
Spacecraft spin inertia = 700 kg · m².
Spacecraft radius   = 1 m.
Yo-yo weights (total)   = 5 kg.

(Ans. $\ell_f = 11.87$ m, $t = 1.89$ s)

27. A dual-spin spacecraft consists of a spinning wheel with spin moment of inertia C_2, and a despun "sail" with spin axis inertia C_1. Total spacecraft (wheel and sail) transverse moment of inertia is A. Show that the equations for wheel spin-up and spacecraft motion are

$$\Omega_3 = (1/C_2) \int T_3\, dt$$

$$\dot{\omega}_1 - (1 - \mu^*)\omega_3\omega_2 + \mu\Omega_3\omega_2 = T_1/A$$

$$\dot{\omega}_2 + (1 - \mu^*)\Omega\omega_1 - \mu\Omega_3\omega_1 = T_2/A$$

where $\mu = C_2/A$, $\mu^* = C_1/A$, ω_1, ω_2, ω_3 are the transverse and spin-up axes angular velocities, and T_1, T_2, T_3 are the transverse and spin-up torques, respectively. What is the spacecraft motion if the spin-up torque is zero?

28. Consider a system of two coaxial cylindrical bodies subject to an external torque T_1 acting on body 1 which is resisted by a bearing friction torque L as shown in Figure A.10.

As a result of the torque T_1 body 1 accelerates at a rate $\dot{r}_1$ and a rate $\dot{r}_2$ where r_1, r_2 are the spin angular velocities, respectively. For spin moments of inertia C_1 and C_2, the equations of motion take the form:

$$C_1\dot{r}_1 = T_1 - L$$

$$C_2\dot{r}_2 = L$$

If the external torque $T_1 = Ke^{-\alpha t}$ where K, α are parameters and if L can be expressed as $L = L_o(r_1 - r_2)$ where L_o is a constant, determine:
(a) Relative angular velocity $\omega = r_1 - r_2$.
(b) Relative displacement $\phi = \int \omega\, dt$.

A.3 Problem Set 3: Control Theory

1. Determine the Laplace transforms of sin ωt and cos ωt.

2. Determine the Laplace transform of $\dfrac{1}{2} at^2$.

(Ans. a/s^3)

3. Solve $\dfrac{d^3y}{dt^3} + 3\dfrac{d^2y}{dt^2} + 10\dfrac{dy}{dt} = 0$

for initial conditions $y(0) = y_o$, $\dfrac{dy(0)}{dt} = \dfrac{d^2y(0)}{dt^2} = \dfrac{d^3y(0)}{dt^3} = 0$. using the Laplace transform method.

$$\left(\text{Ans. } y(s) = \frac{y(O)}{s},\ y(t) = \text{step function}\right)$$

4. Use partial fraction expansions to obtain the inverse Laplace transforms (time functions) of the following polynomials:

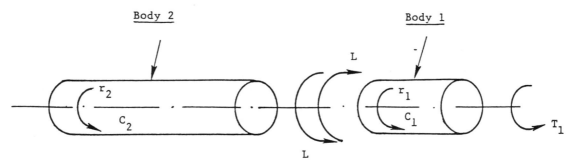

Figure A.10

(a) $\dfrac{A}{(s + a)(s + \sigma)}$ (b) $\dfrac{B}{s(s + b)}$

(Ans. (a) $(A/\sigma - a)(e^{-at} - e^{-\sigma t})$, (b) $(f/b)(1 - e^{-bt})$

5. Find the time response of the integrator in Figure 4.12 for an input of the form $f(t) = e^{-\tau t}$ where τ is a constant.

$$\left(\text{Ans. } \theta_o(t) = \frac{\omega_c}{(\omega_c - \tau)}\, [e^{-\tau t} - e^{-\omega_c t}]\right)$$

6. Derive the time response of modulator error ε in the system shown in Figure A.11 for unit step input θ_i.

7. Reduce the block diagrams and write closed loop transfer functions for:
   (a) Figure A.12.

   (Ans. $KaK/Ts + 1 + KK_f$)

   (b) Figure A.13.

   (Ans. $G_1G_2G_3/[1 + G_2(H_2 + G_1G_3H_1)]$)

8. Assuming zero initial conditions obtain Laplace transformed equations for the system shown in Figure A.14.

9. For the system shown in Figure A.15 find the transfer function $\theta_2(s)/L(s)$.

10. Consider a servo with an open loop pole at the origin of the complex plane and one at -8 rad/s as shown in the block diagram in Figure A.16:
    (a) Plot the closed loop poles as a function of gain k when $k = 7, 15, 16, 17$, and 64. Sketch the root locus and the location of the open loop poles.
    (b) Determine the time response of the servo to a unit step input for $k = 7$. What is the nature of the response when $k > 16$?

11. Draw the Nyquist plot for a damped oscillator system described by a transfer function

$$G(s) = \frac{\omega_n^2}{s^2 + 2\zeta\omega_n s + \omega_n^2}$$

Assume that the damping ratio $\zeta = 0.30$.

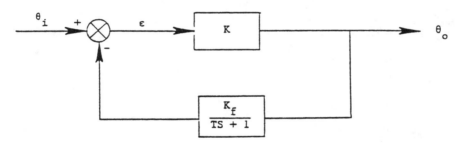

Figure A.11

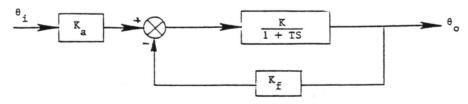

Figure A.12

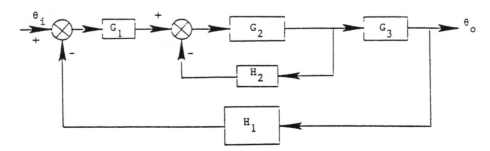

Figure A.13

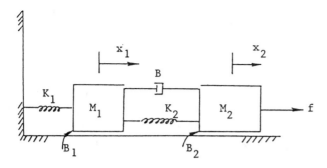

Figure A.14

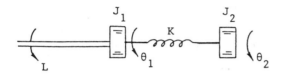

Figure A.15

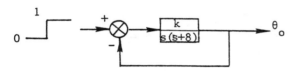

Figure A.16

12. For the open loop transfer function

$$G(s) = \frac{82}{s(1 + s/82)}$$

   (a) Compute open loop frequency response in the range $8.2 < \omega < 147.6$ rad/s.
   (b) Sketch the Nyquist, Bode, and Nichols plots for the frequency range indicated. What is the phase margin for this system?

13. Show that the output of a derived-rate modulator (e.g., Figure 5.9) is approximately proportional to the rate and position of the input signal. Approximate nonlinear element $G(s) = K$ where $K = \infty$ for a step input.

14. Derive equation 5.26 in the text.

15. Three RWs spin about a set of axes aligned with the principal axes of a spacecraft. Let $\vec{h}_o$ be the angular momentum of the spacecraft with the wheels at rest.
   (a) Write a vector expression for the angular momentum $\vec{h}$ of the system if $\vec{h}_i(i = 1 - 3)$ are wheel momenta.
   (b) Express $\vec{h}$ in terms of the body and wheel components if body and angle rates are ω_i and $\Omega_i(i = 1 - 3)$ and the inertias are I_i^o and I^ω_i, respectively.

   (c) Write Euler's equation for the system in body coordinates with external torques T_i applied to the body.
   (d) Write a relationship between $\dot\omega$, $\dot\Omega$, I^ω, and τ^ω where τ^ω is the motor torque applied to the wheels.
   (e) If the wheels are locked, what is the equation of motion of the spacecraft driven by the external torques (gas jets, etc.)?
   (f) If there are no external torques, write the equation of the spacecraft driven by wheel motor torques.

A.4 Problem Set 4: Gravity Gradient, Stability, and Magnetic Stabilization

1. A dumbbell-type spacecraft of negligible central body mass has 10 kg tip masses on 15 meter booms deployed along the local vertical. Compute the gravity-gradient restoring torques per unit angular deviation from the equilibrium orientation. Assume the spacecraft is in a low altitude ($\omega_o \sim 10^{-3}$ rad/s) circular orbit.

2. A space station is in a 500 km circular orbit. It has the following principal moments of inertia:

$$I_{roll} = 9 \times 10^6 \text{ kg} \cdot \text{m}^2$$
$$I_{pitch} = 10^7 \text{ kg} \cdot \text{m}^2$$
$$I_{yaw} = 10^6 \text{ kg} \cdot \text{m}^2$$

   Determine:
   (a) Stable equilibrium orientations in the gravity gradient mode.
   (b) Restoring torques per unit angular deviation. Assume $\omega_o = 1.16 \times 10^{-3}$ rad/s.

   (Ans. -48.4, -32.3, -1.35 N $\cdot$ m/rad)

   (c) Natural frequencies of liberation in terms of orbit rate ω_o.

   (Ans. ω_o, $1.55 \omega_o$, $2 \omega_o$)

3. (a) Derive an expression for the tension in a weightless tether of length L (measured from spacecraft center of gravity) with end mass m_s which is aligned along the local vertical in a circular orbit.

   (Ans. $F = 3LM_s\omega_o^2$)

   (b) Compute the tension in a 100 km weightless tether supporting a 500 kg subsatellite from the Space Shuttle in a low altitude orbit ($\omega_o \sim 1.1 \times 10^{-3}$ rad/s).

   (Ans. $F = 181.5$ N)

   (c) Compute the tether tension due to mass density ρ(mass/unit volume) for a constant cross-section A. Find the maximum tension in the tether.

4. Solve equation 7.20 in the text for pitch libration of a dumbbell satellite in an eccentric orbit neglecting the $\cos \omega_o t$ term.

(a) Find maximum pitch angle θ_2 per percent of orbit eccentricity.

    (Ans. 0.6 deg/% eccentricity)

(b) Discuss the effect of the $\cos \omega_o t$ term on the stability of the solution.

5. Consider a gravity gradient spacecraft configuration with principal moments of inertia I_1, I_2, I_3, a pitch RW of momentum h, and a tip mass (damper) on an extendable boom of length ℓ.
   (a) Write linearized Euler equations of motion and solve for the natural frequencies of the system assuming a rigid boom.
   (b) Discuss the effect of boom flexibility (i.e., first cantilever mode) on the natural frequencies of system.

6. Derive the components of the Earth's magnetic field in spacecraft body coordinates.

7. Compute the average geomagnetic field (magnitude and direction) for a 740 km circular polar orbit which is coplanar with the magnetic pole in the plane of the orbit.

    (Ans. 0.1 gauss)

8. Determine the angular momentum reduction capability for a satellite in the orbit of problem 7 per half orbital revolution if the yaw axis (along the orbit radius) magnetic dipole $M_y = 8,000$ pole · cm ($8ATM^2$) and the pitch axis (along orbit normal) RW angular momentum $h = 20$ N · m · s. Assume that the orbital period of revolution is 100 minutes.

9. Describe a functional block diagram for magnetic momentum unloading in problem 8. Describe a typical magnetic torquer assembly.

10. Compute the spin-axis drift rate for a satellite in a 35,787 km altitude synchronous equatorial orbit with a spin axis (angular momentum) oriented normal to the orbit plane.

Assume:
Spin axis moment of inertia $I_s = 100$ kg · m²
Spin angular velocity $\omega_s = 40$ rpm
Magnetic dipole $M = 1,000$ pole · cm

    (Ans. 4.5×10^{-10} rad/s)

11. Derive the equations of motion for an elastic pendulum (a mass m suspended from a weightless spring of stiffness k) as shown in Figure A.17.
   (a) Solve linearized equations of motion for angle θ and radial excursion r.
   (b) Discuss the effect of the nonlinear coupling terms on the stability of θ motion when the free radial motion is oscillatory with frequency ω_n.

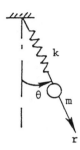

Figure A.17

12. Consider a system of perturbed equations about equilibrium

$$\dot{X} = AX$$

where

$$A = \begin{pmatrix} -3 & -7 \\ 0 & -4 \end{pmatrix}.$$

Determine a Liapunov function of the form

$$V = X'PX$$

where

$$P = \begin{pmatrix} p_{11} & p_{12} \\ p_{12} & p_{22} \end{pmatrix} \quad \text{and} \quad X = \begin{pmatrix} x_1 \\ x_2 \end{pmatrix}$$

from $A'P + PA = -I$ where I is the identity matrix. Test for positive definiteness of P and determine if the solutions about the equilibrium (origin) are asymptotically stable.

    (Ans. $p_{11} = 1/6$, $p_{12} = -1/6$, $p_{22} = 5/12$)

13. Find the region in the y_1, y_2 plane for which the function $V = y_1^2 + y_2^2/(1 + y_2^2)$ defines a family of closed curves $V = c$. What is the condition if V is a Liapunov function for an investigation of stability with respect to y_1 and y_2?

    (Ans. $C \le 1$)

Appendix B

Nomenclature

B.1. Lower and Uppercase Symbols

A, a —area or acceleration

$I, A, B, C,$ —moments of inertia

c —velocity of light

$E_\alpha, e_\alpha \ (\alpha = 1 - 3)$ —reference and body-fixed coordinates respectively

H, h —angular momentum

R, r —radius

T —torque or kinetic energy

q_i —quaternions or generalized coordinates

ℓ —moment arm

M, m —mass

P —period or pressure

g —gravitational constant

p_i, s_i —root i

s —Laplace variable

V, v —potential energy or velocity

$x, y, z,$ —coordinates, variables

F —force

W —work

I_{sp} —specific impulse

B.2 Greek Symbols

$\alpha, \beta, \gamma, \delta, \mu, \eta$ —subscripts or angles

θ, ϕ, ψ —Euler angles

λ, ω —frequency or angular rate

τ, ζ —time constant and damping ratio, respectively

μ —gravitational constant for Earth or inertia ratio C/A, angle

Γ —gradient matrix

B.3 Other Symbols

$(\rightarrow)(\hat{\ })$ —vector and unit vector, respectively

$(\bar{\bar{\ }})$ —dyadic

$(\ )$ —column matrix

$| \ |$ —absolute value or matrix

$(\cdot)$ —velocity (time derivative of position)

$(\ddot{\ })$ —acceleration (time derivative of velocity)

$(')$ —transpose of a matrix

Index